Samuel C. Perry
Electrochemical Carbon Dioxide Reduction

Also of Interest

Green Chemistry and Technology.
In: Green Chemical Processing, 6
Mark Anthony Benvenuto and George Ruger *(Eds.), 2021*
ISBN 978-3-11-066991-6, e-ISBN 978-3-11-066998-5

Sustainable Process Engineering
Gyorgy Szekely, 2021
ISBN 978-3-11-071712-9, e-ISBN 978-3-11-071713-6

Sustainable Utility Systems.
Modelling and Optimisation
Petar Sabev Varbanov, Jiří Škorpík, Jiří Pospíšil and Jiří Jaromír
Klemeš, 2020
ISBN 978-3-11-063004-6, e-ISBN 978-3-11-063009-1

Carbon Dioxide Utilization.
Volume 1: Fundamentals
Michael North and Peter Styring (Eds.), 2019
ISBN 978-3-11-056309-2, e-ISBN 978-3-11-056319-1

Product and Process Design.
Driving Innovation
Jan Harmsen, André B. de Haan and Pieter L. J. Swinkels, *2018*
ISBN 978-3-11-046772-7, e-ISBN 978-3-11-046774-1

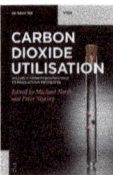

Samuel C. Perry

Electrochemical Carbon Dioxide Reduction

—

DE GRUYTER

Author
Dr. Samuel C. Perry
Faculty of Engineering and Physical Sciences
University of Southampton
University Road
Southampton SO17 1BJ
Great Britain
Email: s.c.perry@soton.ac.uk

ISBN 978-1-5015-2213-0
e-ISBN (PDF) 978-1-5015-2223-9
e-ISBN (EPUB) 978-1-5015-1544-6

Library of Congress Control Number: 2021936904

Bibliographic information published by the Deutsche Nationalbibliothek
The Deutsche Nationalbibliothek lists this publication in the Deutsche Nationalbibliografie;
detailed bibliographic data are available on the Internet at http://dnb.dnb.de.

© 2021 Walter de Gruyter GmbH, Berlin/Boston
Cover image: Olivier Le Moal/iStock/Getty Images Plus
Typesetting: Integra Software Services Pvt. Ltd.
Printing and binding: CPI books GmbH, Leck

www.degruyter.com

Contents

1 Introduction to electrochemical carbon dioxide reduction

1.1 The role of electrochemical CO_2 reduction in global decarbonisation

The threat of rising CO_2 levels and resultant global warming is arguably one of the most important challenges faced by researchers today. The global, societal relevance of this issue is demonstrated by the large number of governmental and industrial actions that have been established, such as the Intergovernmental Panel on Climate Change, Global Climate Change Initiative and the European Strategic Energy Technology Plan [1]. Ambitious global targets have been set to provide a global net-zero carbon emissions by the year 2050, an essential target to meet if the global temperature rise is to be limited to the target 1.5 °C [2]. To this end, we have seen promising investments in green technologies and practices. Despite this, the current global trend in CO_2 emissions is not on track to meet this 2050 target [3], despite the clear public support for environmentally minded practices. In fact, global CO_2 levels are still continually rising by around 2 ppm each year, recently exceeding 400 ppm for the first time [4].

Many sources of anthropogenic carbon emissions are intuitive and have been within the scope of public attention for a number of years. Primary emission sources such as electricity production and transport networks have led to increasing investments in renewable energy sources (solar, wind, hydroelectric, etc.) and green transport solutions (whole and hybrid electric motors). Within building and industrial processes, a major source of CO_2 emission is from energy consumption. However, several key processes produce CO_2 directly as a by-product of a chemical reaction, such as steel manufacture and ammonia synthesis. Across all sectors, waste disposal by incineration is also a key contributor. Sources of anthropogenic CO_2 and the fate of CO_2 emissions within the carbon cycle are detailed in Table 1.1.

Much discussion surrounding CO_2 challenges focusses on the decarbonisation of these key processes, especially in energy and transport sectors. Increased research and commercial uptake into green technologies has lowered the cost of renewable energies through solar and wind. Further developments in this area have worked towards enhanced energy storage solutions to offset challenges of intermittent energy production through green energy sources, such as through batteries, redox flow batteries, pumped hydroelectric power or biofuel production [5].

Alongside decarbonisation initiatives, several large-scale carbon sequestration projects have been implemented. Rather than lowering the rate of CO_2 production, these sites aim to capture CO_2 in some benign form that can be easily stored, rather than released into the environment. As of 2013, there are 18 large-scale carbon capture

https://doi.org/10.1515/9781501522239-001

Table 1.1: Global carbon fluxes given in gigatonnes of carbon per year (GtC year^{-1}).

Carbon flux	Amount (GtC year^{-1})	Year	Citation
Carbon cycle overview			
Air–land natural exchange	120	2007	[7]
Air–ocean natural exchange	90	2007	[7]
Anthropogenic CO_2 emissions	11	2014	[8]
Anthropogenic greenhouse gas emission by CO_2 equivalent	13	2010	[9]
Fate of anthropogenic carbon emissions			
Net ocean uptake	4	2014	[8]
Net land uptake	3	2014	[8]
Accumulated in atmosphere	4	2014	[8]
Anthropogenic carbon emissions by sector			
Electricity generation	3.3	2010	[9]
Agriculture and deforestation	3.2	2010	[9]
Industry	2.8	2010	[9]
Transport	1.9	2010	[9]
Buildings	0.9	2010	[9]
Other	1.4	2010	[9]

Values are accurate as of the given years.
Adapted with permission from reference [6], Copyright © 2020 American Chemical Society.

projects that are actively removing around 9.6 Mt of carbon every year. These units are hyphenated with CO_2 sources such as industrial processes or biomass electrical plants, so that CO_2 is captured and stored at the point of production [10]. Although just a small fraction of global CO_2 production (note that Table 1.1 features carbon on the order of gigatonnes), there are promising opportunities for further upscaling and integration with additional CO_2 sources.

What is clear, based on global CO_2 production and the uptake of new technologies, is that there cannot be one simple solution to tackling the carbon challenge. In order to meet the 2050 emission target, additional decarbonisation technologies must be integrated into the energy, transport and manufacture sectors. One such technology that is receiving growing interest is the electrochemical reduction of CO_2 into value-added products. The core concept is the reaction of CO_2 at an electrode surface via the CO_2 reduction reaction (CO_2RR), where a series of electron and proton transfers drives the formation of the desired product. This leads to CO_2 acting as a carbon source for the production of a new material, either for the production of industrially relevant materials or for energy storage applications.

It is worth mentioning at this point that none of the global CO_2RR literature proposes atmospheric CO_2 as a viable carbon source for electrochemical synthesis. The quantity of CO_2 emitted globally is simply too large to offset using electrochemical techniques. This point was demonstrated in a review by Pletcher, who showed back-of-the-envelope calculations for the power requirements to use atmospheric CO_2 as a direct feed gas for a CO_2 electrolyser. Even making generous assumptions about the reactor performance and charge efficiency, conversion of the annual globally produced CO_2 volume would require 140×10^{15} Wh year^{-1}. When we consider that the global energy production is only 23×10^{15} Wh year^{-1}, there is clearly no achievable future in this approach [11].

The practical applications of CO_2RR technologies versus decarbonisation approaches can be identified through the varied carbon sources displayed in Table 1.1. When thinking about carbon produced from transport, clearly the collection and storage or direct electrolysis of CO_2 from every fossil fuel engine is not a practical solution, so decarbonised solutions such as electric- or hydrogen-powered vehicles are the focus of current research efforts. Conversely, industrial processes such as fermentation or certain chemical syntheses cannot occur without the formation of substantial amounts of CO_2, making hyphenation of these processes with CO_2 electrolysers a practical solution. It is also important to consider the current source of potential CO_2RR products. Many of the CO_2RR products are presently sourced from fossil fuels and are produced in sizeable quantities. Switching these production routes to a CO_2 feedstock would provide a sizeable reduction in fossil fuel consumption, which comes with significant environmental benefits even before we consider the impact of the removed CO_2.

This demonstrates how the CO_2RR fits into its niche within the global effort to lower global CO_2 levels. Areas that can be decarbonised will be targeted with new green technological solutions focussing on renewable energy capture and storage. Areas that cannot be decarbonised can be integrated with CO_2 electrolysers so that CO_2 is captured and converted, rather than released into the atmosphere (Figure 1.1).

As a practical limitation for CO_2 electrolysers, the hyphenation between CO_2 source and electrolyser will not be direct; an electrolyser plugged into an industrial exhaust flue will, in all likelihood, quickly become poisoned either through chemical poising of the catalyst or through gas channels within the electrolyser becoming blocked by particulates. Instead, the goal should be to capture and purify the CO_2 onsite, then transport CO_2 before feeding it into the electrolyser. Certain CO_2 sources are immediately more desirable for this application based on the inherent cleanliness of the CO_2 source. For example, CO_2 from fermentation processes requires relatively little cleaning before it can be used in a reactor, whereas CO_2 from coal-burning power stations contains significant quantities of nitrous and sulphurous oxides that would have to be removed.

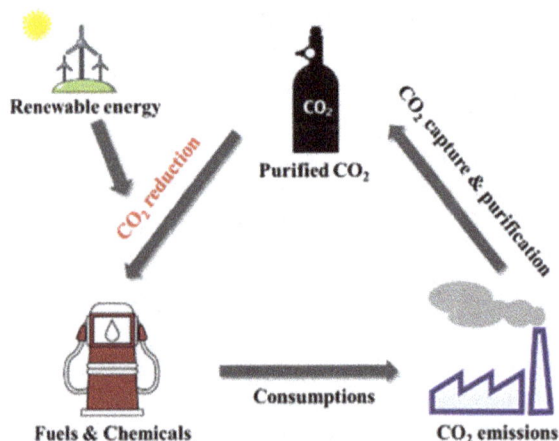

Figure 1.1: Schematic for a simplified anthropogenic carbon cycle. CO_2 emissions are captured and purified, then converted into fuels and chemicals using a renewable energy source. The real system would, of course, be far more complicated, but the aims and motivations will remain the same. Reproduced with permission from reference [12], Copyright © 2020 X. Zhang et al. Published by Elsevier Ltd.

1.2 CO_2 reduction to value-added materials

CO_2 can be simply thought of as oxidised carbon. In this case, any conversion to alternative carbonaceous species must involve a reduction of the carbon oxidation state. Broadly this is referred to as CO_2 reduction, although there are a number of subsets depending on how this reduction in oxidation state is achieved; CO_2 hydrogenation uses a thermal reaction with hydrogen, CO_2 fixation uses biological or biologically inspired process akin to photosynthesis and CO_2 electrochemical reduction uses charge transfer at an electrode. The CO_2RR can be thought of as a power-to-gas or power-to-liquid process. The basic reactor is fed with water at the anode and CO_2 at the cathode, resulting in O_2 evolution and CO_2 reduction, respectively. The corresponding half-cell reactions and overall reaction are given in the following equations:

$$a CO_2 + n H^+ + n e^- \rightarrow CO_2RR \text{ product} + b H_2O \tag{1.1}$$

$$2H_2O \rightarrow O_2 + 4H^+ + 4e^- \tag{1.2}$$

$$a CO_2 + c H_2O \rightarrow CO_2RR \text{ product} + d O_2 \tag{1.3}$$

CO_2 reduction is relatively unusual in that there is a broad scope of different materials that can be produced depending on the precise reaction conditions. This gives CO_2 reduction an inherently broad scope of potential applications, thanks to the range of products available. Importantly, each of these products has a sizeable global market, and so all of the products here discussed are potential targets for CO_2RR technologies (Figure 1.2).

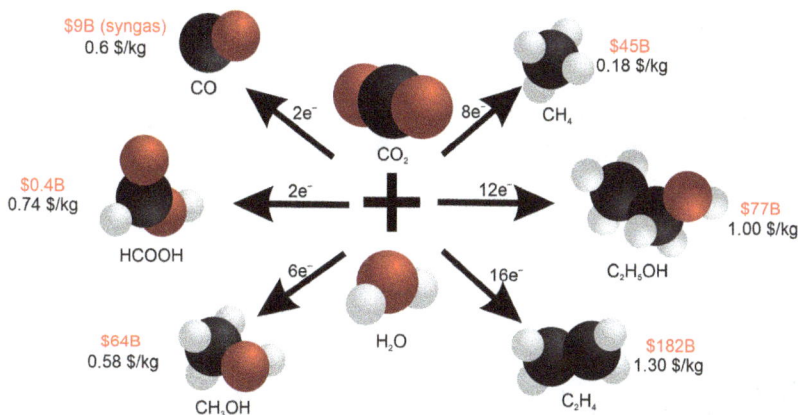

Figure 1.2: Viable targets for CO_2RR reactors on a larger scale. Given values show the market size (red), unit price (black) and number of electrons transferred in the electrochemical reaction (next to the arrows). Adapted with permission from reference [13], available under Creative Commons (CC BY 4.0) license, Copyright 2019 Song et al.

1.2.1 Carbon monoxide and syngas

The first product to discuss is carbon monoxide. This is arguably the simplest CO_2RR product to produce, since the mechanism requires only two electron transfers:

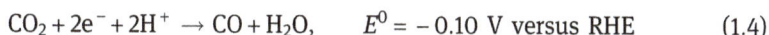

$$CO_2 + 2e^- + 2H^+ \rightarrow CO + H_2O, \qquad E^0 = -0.10 \text{ V versus RHE} \qquad (1.4)$$

Currently, CO production at large-scale plants uses coal gasification, steam reformation of natural gas or the oxidation of hydrocarbons, all of which come with a clear and sizeable carbon footprint [14]. CO has a number of uses and applications in chemical synthesis on a large industrial scale. For example, CO conversion to formic acid via the hydrolysis of methyl formate accounts for 450,000 tonnes of CO utilisation every year [15]. Other possible products include acetic acid, acrylic acid, propanoic acid, acetic anhydride, dimethylformamide and phosgene [16]. Outside of chemical synthesis, CO is used in metal manufacture to remove impurities from blast furnaces and in nickel, iron, cobalt, chromium, tungsten and molybdenum production [17].

Many CO applications come from syngas, a mixture of CO and H_2 gases (Figure 1.3). The precise H_2:CO ratio varies depending on the application, with several possible applications ranging from chemical synthesis to biofuel production and energy generation [18]. Further uses become apparent when considering subsequent application of syngas products, such as using ammonia in fertiliser manufacture or methanol for liquid fuel for formaldehyde production. This is of particular interest to electrochemical

CO_2RR designs, since H_2 production via proton reduction occurs over the same potentials as CO_2 reduction to CO:

$$2H^+ + 2e^- \rightarrow H_2, \quad E^0 = 0.00 \text{ V versus RHE} \tag{1.5}$$

This presents the possibility to produce a $CO:H_2$ gas mixture directly from a CO_2RR reactor. Through a customised reactor design and applied potential, it could then be possible to generate a customised $CO:H_2$ gas mixture directly for a targeted application.

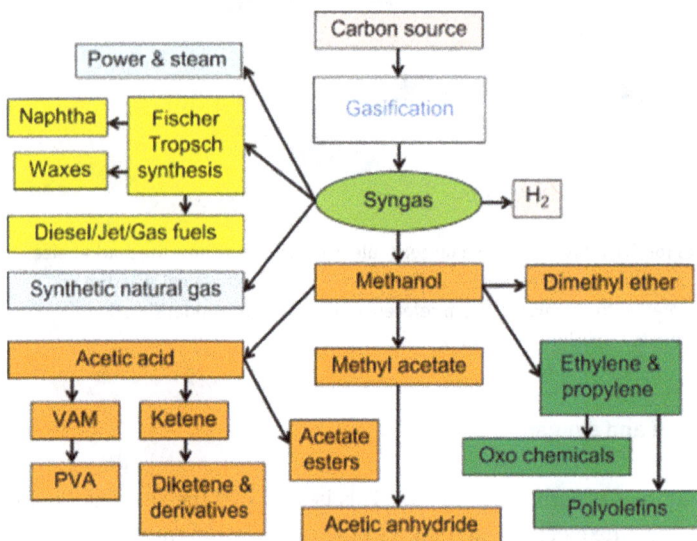

Figure 1.3: A number of syngas derivatives along with their synthetic routes. Reproduced with permission from reference [19], Copyright © 2017 Elsevier Ltd.

1.2.2 Formic acid

Formic acid (HCO_2H) is one of the most widely researched targets for the CO_2RR, thanks to its wide variety of uses across chemical industry and in search for new biofuels. As a marketable chemical, formic acid has applications as a preservative and as an antibacterial agent 1. It can also substitute a number of inorganic acids in chemical syntheses and industrial processes, with that advantage that formic acid is less corrosive than the alternatives and does not lead to nitrate, phosphate or sulphate loading in wastewater [20].

Formic acid also has potential applications in a number of chemical synthesis reactions. Formate dehydration over solid-acid catalysts liberates CO, which can then be used as a C_1 building block for a range of chemical syntheses [21, 22]. Alternatively, formic acid can be used as a hydrogen donor for chemical reactions that would normally use H$_2$ gas over a heterogeneous catalyst, such as hydrogenation or deoxygenation reactions:

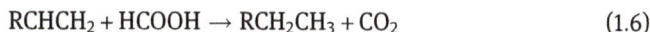

$$RCHCH_2 + HCOOH \rightarrow RCH_2CH_3 + CO_2 \tag{1.6}$$

$$RO + HCOOH \rightarrow R + H_2O + CO_2 \tag{1.7}$$

Current worldwide demand for formic acid is around 950,000 tonnes per year, with usage expected to grow as more industries adopt it as a relatively non-toxic and non-corrosive acid, or as an alternative to compressed H$_2$ gas [23]. Possibly the most widely researched application for formic acid is as a hydrogen source for fuel cell technologies. Formic acid exists in an equilibrium between H$_2$ and CO$_2$ gases:

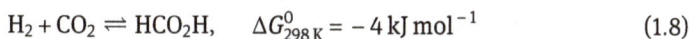

$$H_2 + CO_2 \rightleftharpoons HCO_2H, \quad \Delta G^0_{298\,K} = -4\,kJ\,mol^{-1} \tag{1.8}$$

Formic acid can therefore be thought of as a liquid-phase hydrogen storage, where reversal of the equilibrium produces H$_2$ for power generation (Figure 1.4). Such liquid-phase H$_2$ storage is greatly desirable as it removes the cost and hazards associated with storing gas-phase H$_2$ in high-pressure cylinders. As a comparison, 1 L of liquid formic acid contains 26.5 M (53 g L^{-1}) of H$_2$, whereas 1 L of compressed H$_2$ gas at a high pressure of 22 MPa contains only 9.8 M H$_2$ [20]. Since the liberation of H$_2$ from formic acid returns our starting CO$_2$ material, it is possible to envisage a catalytic loop where CO$_2$ produced from formic acid decomposition is captured and reduced to restore the formic acid supply. This is desirable, since the evolved CO$_2$ from formic acid decomposition should be highly pure. CO$_2$ would therefore only require separation from H$_2$, which will need to be done anyway to access the H$_2$ fuel.

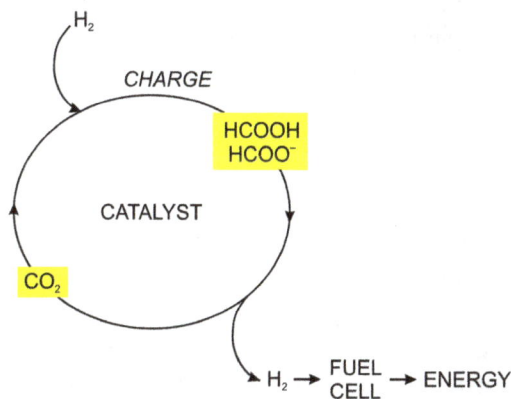

Figure 1.4: Schematic catalytic loop for hydrogen storage in formic acid. Formic acid decomposition produces H$_2$ and CO$_2$. H$_2$ is fed into the fuel cell as a power source. CO$_2$ is fed into a CO$_2$RR reactor to reproduce formate as a H$_2$ storage option. Adapted with permission from reference [20], Copyright ©2018 Wiley-VCH Verlag GmbH & Co. KGaA, Weinheim.

1.2.3 Methane and methanol

Methane and methanol are two of the higher order C_1 CO_2RR products, requiring eight and ten electron transfers, respectively, to give the end product. Methane is a well-known fuel source. As a primary component of natural gas, synthetic methane could be employed in any system that currently uses natural gas as a primary fuel [24]. This is a sizeable market, with global natural gas usage of up to 3,986 billion cubic metres. A further advantage is that, since natural gas usage is already widespread, CO_2RR-generated methane could be rapidly integrated into existing infrastructure.

Similarly, methanol is a promising option for CO_2RR-generated fuel, thanks to its low freezing point, high energy content and importantly because it can be used in standard combustion engines with minimal modifications [25]. Methanol fuel can also be used for energy generation via the direct methanol fuel cell (DMFC). As a liquid fuel, methanol offers a greater volumetric energy density than hydrogen and can be simply stored in plastic tanks. DMFCs are attracting increasing interest with continually improving efficiencies and energy outputs, although challenges surrounding anode activity and methanol crossover still need to be solved before widespread adoption can be realised [26]. Current methanol consumption sits at 100 million tonnes per year [27], which would be expected to greatly increase as methanol uptake as a viable biofuel increases.

At first glance, methane and methanol combustion may seem like a continued reliance on CO_2-producing fuel sources. However, the combustion of these products from CO_2RR is retuning CO_2 to the atmosphere after it was previously captured, which is less harmful than the combustion of natural gas or gasoline, which releases CO_2 that was previously trapped underground. Electrochemically generated fuels would also have a greatly improved purity versus fossil fuels, providing energy with greatly reduced NO_x and SO_x pollutant by-products.

1.2.4 Ethylene and ethanol

Ethylene and ethanol make up the most commonly produced C_2 CO_2RR products. Ethylene's dominant use is in production of polyethylene. It is also used in the manufacture of detergents and synthetic lubricants [28]. Simple chemical conversion opens up a wider range of applications. Oxidation gives ethylene oxide, which is a precursor to ethylene glycol and polyester. Alkylation can give ethylbenzene, which is a precursor for styrene and polystyrene production [29]. These widespread industrial uses lead to a global ethylene usage of around 158 million tons, which is expected to grow by a further 4.5% by 2027, making it one of the largest potential CO_2RR product markets [30].

Ethanol also has uses in a range of chemical syntheses for the production of a number of glycol ethers and amines. Ethanol dehydration is also a feasible means for ethylene production at an industrial scale, which is growing in interest thanks to expansion of bioethanol production from biomass fermentation, which produces over 100 billion litres of ethanol a year [28, 31]. Ethanol also has applications as a biofuel, either in direct ethanol fuel cells or mixed with gasoline for transport applications [32].

1.2.5 Additional products

One of the distinguishing features of CO_2RR is its ability to produce a wide range of varying products depending on the precise reaction conditions. The previous sections covered CO_2RR products that arguably attract the most interest, thanks to a combination of economic interest and the relative simplicity of producing a selective system with currently available technologies. A number of other species have also been detected during CO_2RR, although most of these are mentioned as by-products in search of a different, previously mentioned material, rather than as the subject of its own work.

Of course, there are a number of exceptions to this that are worth mentioning. Organic acids such as acetic and oxalic acid are regularly detected in liquid-phase CO_2RR, sometimes with relatively high efficiencies. Acetic acid is primarily used in the manufacture of vinyl acetate monomer, which is then used in the production of polyvinyl acetate, which is used in paints and adhesives. Other applications for acetic acid or vinyl acetate monomer extend to ink manufacture and as a chemical solvent, with a global demand of over 13 million tonnes [33]. Oxalic acid also has applications in a number of industrial areas, such as in textile manufacture, in metal surface treatments and in the separation of rare earth materials [34], leading to a global usage of around 120,000 tonnes [35].

Other products that are often detected include C_{2+} hydrocarbons and oxygenates such as ethane, ethylene glycol, n-propanol and n-butane. However, these products are presently detected in significantly lower yields than other available options, and so it is premature to discuss these as viable CO_2RR products from an economic perspective.

1.3 Primary challenges

1.3.1 Single product selectivity

On the face of it, the usual ability of CO_2 reduction to simultaneously produce multiple reaction products in relatively high yields may seem advantageous. However, this presents a key challenge in the economic viability of the process, the ability to selectively produce just one of the many possible products. The challenge is highlighted by the standard potentials (E^0) for the CO_2RRs to a number of target products [36]:

$$CO_2 + 2e^- + H^+ \rightleftharpoons HCO_2^-, \qquad E^0 = -0.02 \text{ V versus RHE} \qquad (1.9)$$

$$CO_2 + 2e^- + 2H^+ \rightleftharpoons CO + H_2O, \qquad E^0 = -0.10 \text{ V versus RHE} \qquad (1.10)$$

$$CO_2 + 6e^- + 6H^+ \rightleftharpoons CH_3OH + H_2O, \qquad E^0 = 0.10 \text{ V versus RHE} \qquad (1.11)$$

$$CO_2 + 8e^- + 8H^+ \rightleftharpoons CH_4 + 2H_2O, \qquad E^0 = 0.03 \text{ V versus RHE} \qquad (1.12)$$

$$2CO_2 + 8e^- + 7H^+ \rightleftharpoons CH_3COO^- + 2H_2O, \qquad E^0 = -0.26 \text{ V versus RHE} \qquad (1.13)$$

$$2CO_2 + 10e^- + 10H^+ \rightleftharpoons CH_3COH + 3H_2O, \qquad E^0 = 0.05 \text{ V versus RHE} \qquad (1.14)$$

$$2CO_2 + 12e^- + 12H^+ \rightleftharpoons CH_3CH_2OH + 3H_2O, \qquad E^0 = 0.09 \text{ V versus RHE} \qquad (1.15)$$

$$2CO_2 + 12e^- + 12H^+ \rightleftharpoons C_2H_4 + 4H_2O, \qquad E^0 = 0.08 \text{ V versus RHE} \qquad (1.16)$$

$$2CO_2 + 12e^- + 12H^+ \rightleftharpoons CH_3CH_2OH + 3H_2O, \qquad E^0 = 0.09 \text{ V versus RHE} \qquad (1.17)$$

$$3CO_2 + 16e^- + 16H^+ \rightleftharpoons CH_3CHCH_2OH + 5H_2O, \qquad E^0 = 0.11 \text{ V versus RHE} \qquad (1.18)$$

$$3CO_2 + 18e^- + 18H^+ \rightleftharpoons CH_3CH_2CH_2OH + 5H_2O, \qquad E^0 = 0.21 \text{ V versus RHE} \qquad (1.19)$$

In most electrochemical systems, a fair degree of selectivity is accessible through the selection of an appropriate applied potential (E). For a given reduction reaction, E must be sufficiently negative versus E^0 to drive the reduction reaction. The potential difference between E and E^0 is defined as the overpotential according to the following equation:

$$\eta = E - E^0 \qquad (1.20)$$

In the case of CO_2 reduction, the sizeable overlap in E^0 means that any overpotential that is sufficiently large to form one product will produce any one of a number of others. It is not surprising that CO_2 reduction reactors invariably produce a mixture of product gases. At this point, it is worth mentioning that eqs. (1.9)–(1.19) show the standard potentials (E^0) as calculated from thermodynamic data, that is, from the Gibbs free energies of the corresponding reactions. However, real experimental standard potentials ($E^{0'}$) are found to be considerably more negative than these thermodynamic values. This is due to the key rate-determining step in the CO_2RR – the first electron transfer to give the CO_2^- radical anion (see Section 2.1).

A further complication arises when considering the electrochemistry of water. The solvent window of water is defined by the onset of cathodic and anodic electrolyses at negative and positive potentials, respectively:

$$2H_2O + 2e^- \rightleftharpoons H_2 + 2OH^-, \quad E^0 = 0.00 \text{ V versus RHE} \qquad (1.21)$$

$$2H_2O \rightleftharpoons O_2 + 4e^- + 4H^+, \quad E^0 = 1.23 \text{ V versus RHE} \qquad (1.22)$$

When we compare the standard potentials for water electrolysis to those of the CO_2RR, we see that any applied potential that is negative enough to drive CO_2 reduction will inevitably result in some degree of hydrogen evolution reaction (HER). It is worth noting that this comparison is only to the thermodynamically calculated E^0. In practice, the experimental $E^{0'}$ is considerably more negative than the theoretical values due to the very negative potential for the first electron transfer to the CO_2 radical anion, $E^0 = -1.85$ V versus RHE (eq. (2.1)). A further problem in real electrochemical cells is due to the low solubility of CO_2 in aqueous solutions [36]. The electrode–electrolyte interface will experience low concentrations, and slow diffusion rates, of CO_2 compared to an abundance of electrochemically available H_2O.

This is not entirely prohibitive to CO_2RR. Strongly adsorbing CO_2, and its reaction intermediate CO, can successfully hinder the HER by blocking active sites for water reduction [37]. However, as the overpotential is further increased, CO_2 is more rapidly consumed and the reduction rate becomes limited by the rate of CO_2 diffusion to the electrode surface. H_2O in aqueous electrolyte is abundantly available, and so it has no such transport limit, so that the overpotential increases, and the rate of H_2 evolution by water electrolysis also increases. Unless significant chemical and engineering interventions are made, CO_2RR reactors at high overpotentials will produce large quantities of H_2 and little else.

The desire for selective production of a single product stems from two primary motivations: (i) charge efficiency and (ii) reduced downstream costs. The challenge of charge efficiency is related to energy costs; a cell operating at 80% efficiency means that 80% of the charge pass at the electrode is used to produce the desired CO_2RR product but is wasting 20% of the charge passed. This 20% could be going towards any of the alternative CO_2RR or could be lost due to background electrochemical processes. This is usually described in terms of the faradaic efficiency (FE), defined as the ratio of charge used for a specific product ($Q_{product}$) versus the total charge passed (Q_{total}):

$$FE = \left(\frac{Q_{product}}{Q_{total}} \right) \times 100\% \tag{1.23}$$

The issue of downstream costs becomes apparent when we consider what happens after our reactor has completed its operation. The outflow gas has been captured and stored and now contains a mixture of gases. In order to access our desired product, we now need to pass this gas through a separation and purification stage, which comes with a sizeable additional cost. It is worth mentioning at this point that a separation and purification stage is unlikely to be removed entirely from a commercial CO_2RR reactor. A 100% selective reactor is highly unlikely, and even in reactors with excellent selectivity we must also consider the need to recover the product from unreacted CO_2 in the outflow. Despite this, keeping the outflow gas as pure as possible can minimise these costs, which greatly increases the viability of the technology.

1.3.2 Changes to the solution environment

When considering the reaction environment in a CO_2RR reactor, it is vital to consider not only the consumption of CO_2 and generation of product but also the wide range of alternative products and intermediates involved in the possible side reactions over the same potential range. The first consideration is that CO_2 in aqueous media is not benign but exists as in equilibrium with aqueous bicarbonate and carbonate ions [38]:

$$CO_2 + H_2O \rightleftharpoons HCO_3^- + H^+ \quad (pK_a = 7.8) \tag{1.24}$$

$$HCO_3^- \rightleftharpoons CO_3^{2-} + H^+ \quad (pK_a = 10.3) \tag{1.25}$$

It is clear that an electrolyte that is bubbled with CO_2 cannot be assumed to simply contain $CO_{2(aq)}$, rather the CO_2 speciation will be strongly pH dependent, as defined by the equilibria in eqs. (1.24) and (1.25). The nature of dissolved CO_2 is sometimes discussed in terms of the total dissolved inorganic carbon (DIC), which comprises the total CO_2, HCO_3^- and CO_3^{2-}. For completeness, CO_2 dissolution would proceed first via the formation of H_2CO_3, which then takes part in the corresponding proton exchange equilibria:

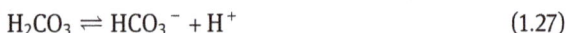

$$CO_2 + H_2O \rightleftharpoons H_2CO_3 \tag{1.26}$$

$$H_2CO_3 \rightleftharpoons HCO_3^- + H^+ \tag{1.27}$$

However, since H_2CO_3 has been found to make up < 0.01% of the total DIC, it is usually omitted from CO_2-phase discussions [39]. It is especially important to consider these equilibria when carrying out CO_2 electrolysis over extended periods of time. This was well demonstrated by Zhong et al., who measured the change in electrolyte pH after continued CO_2 bubbling (Figure 1.5).

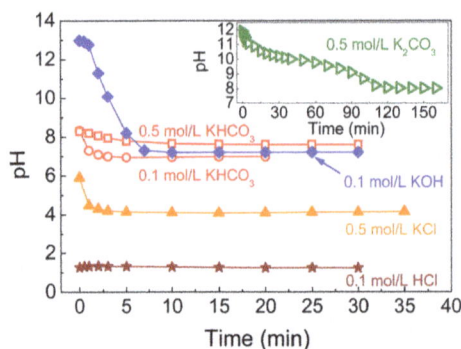

Figure 1.5: pH as functions of time in different solutions while bubbling with CO_2 at a flow rate of 50 mL min^{-1}. Reproduced with permission from reference [40], Copyright © 2014 American Chemical Society.

As CO_2 is continually bubbled through solution, the $CO_2/HCO_3^-/CO_3^{2-}$ equilibrium establishes itself. This leads to a rapid alteration in the solution pH, with a sizeable shift after only a few minutes. This is most noticeable in non-carbonate-based electrolytes. About 0.1 M $KHCO_3$ showed a pH shift from ~ 8.5 to ~ 7, whereas 0.1 M KOH showed a much larger decrease from pH 13 to pH ~ 7. Increasing the concentration of $KHCO_3$ gives a smaller shift in solution pH, as the CO_2 bubbling has less of an impact on the total DIC.

Clearly, when constructing a reactor for CO_2RR, it is vital to understand that the working electrolyte may not be the same as its starting condition. Even if the bulk solution is not altered, continued CO_2 flow and consumption at the electrode will give a dynamic local reaction environment that can have a significant impact on the rate of CO_2RR and the selectivity towards a specific product. That is not to say that the concentrated carbonate is the only option for CO_2RR reactors. Many leading designs use hydroxide- or halide-based electrolytes, as will be later discussed. The emphasis is that a full consideration is needed for both the starting conditions versus the continually evolving conditions after extended operations.

The changing speciation of DIC is best represented by the Pourbaix diagram for CO_2 in aqueous media (Figure 1.6).

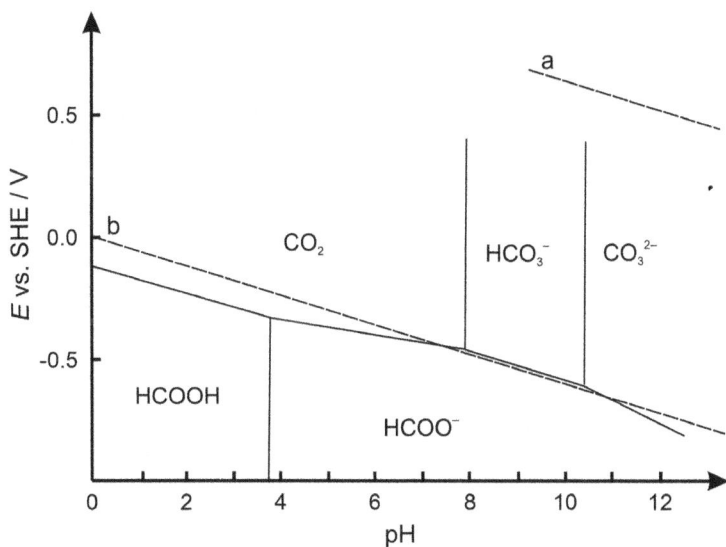

Figure 1.6: Pourbaix diagram for the carbon–water system at 298 K, CO_2 partial pressure $(P_{CO_2}) = 10^5$ Pa. Adapted with permission from reference [41], Copyright © 2008, Springer Science.

The Pourbaix diagram is a plot of E versus pH, which serves as a phase diagram for CO_2 in aqueous media. The well-defined sections display what phase of CO_2 is the most thermodynamically favoured under any given combination of E and pH. The boundaries are defined as the points where activities of neighbouring species are equal. Reading along the x-axis from left to right shows phase changes as the environment becomes more alkaline, and reading along the y-axis from top to bottom shows phase changes as the environment becomes more reducing. In this way, as well as the changes of equilibrium species between CO_2, HCO_3^- and CO_3^{2-}, we can also see the electrochemical transition between CO_2 and its simplest aqueous reduction product, formate (eq. (1.9)). Similar diagrams could, of course, be produced for all possible CO_2RR products.

The long-dashed lines show the solvent window. Line "a" represents the upper limit due to water oxidation to O_2 (eq. (1.22)) and line "b" represents the lower limit due to water reduction to H_2 (eq. (1.21)). The negative gradient of these lines demonstrates the pH dependence of the oxygen evolution reaction and HER, as the pH decreases the standard potential becomes less negative. The same negative gradient can be observed between CO_2 and the reduced formate. Low pH solutions can therefore facilitate both the CO_2RR and HER.

The impact of electrolyte pH is not only important for the starting solution, it is also essential to consider the local reaction environment close to the electrode surface. Sticking with the two-electron reduction of CO_2 to formate, it can be seen from eq. (1.9) that CO_2RR by this mechanism results in a stoichiometric amount of H^+ being consumed per formate produced (in high pH solutions, water is consumed instead, leading to a stoichiometric production of OH^-). In either case, CO_2RR causes the localised environment to become progressively more alkaline after continued electrolysis. Combining this with the observations from Figure 1.5 gives competing impacts on the solution pH; CO_2 bubbling acts to decrease, while continued CO_2RR increases the pH. This makes it challenging to predict the precise CO_2RR environment, adding complications to catalyst design and especially to computational modelling.

It is important to note that the dissolution of CO_2 into the bicarbonate/carbonate equilibria can occur at such a scale that the impact can be seen in the volumetric CO_2 flow. On a small scale, this is an intuitive impact. Ignoring the impact of electrochemical consumption, a set volume of CO_2 enters the reactor, a certain percentage dissolves in the electrolyte; therefore, a smaller volume leaves the reactor. This should be considered when investigating the products from CO_2RR reactors, since the calculated percentage distribution will differ if it is calculated based on the inlet or outlet volumetric flow rate. This is particularly important when operating at high current densities, where CO_2 dissolution and consumption combine to give a sizeable volumetric change.

1.3.3 Carbon costs

When considering the environmental benefits of the CO_2RR, it is important to consider the whole carbon cycle for the present manufacturing process, not only what is occurring within the reactor. Any CO₂RR-based technology will contain multiple components, each with specific energy and operation requirements. First, considering CO_2RR to value-added chemical products, broader applications must consider specific units for CO_2 capture, purification and concentration, CO_2 electrolysis and product separation (Figure 1.7).

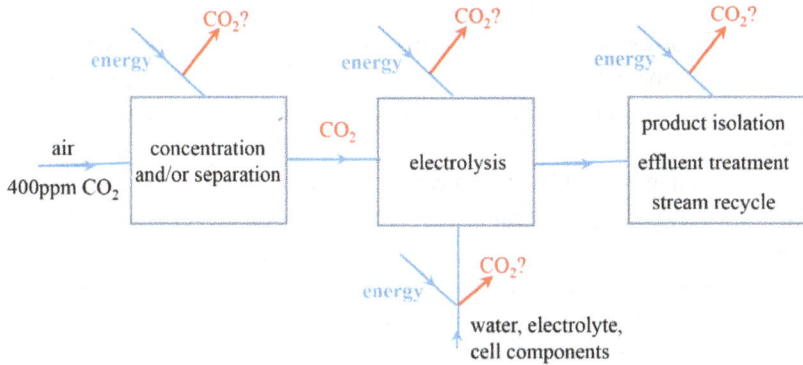

Figure 1.7: Scheme for a plant to remove CO_2 from the atmosphere using a CO_2 electrolyser. Energy inputs and potential sources of CO_2 production are highlighted by blue and red arrows, respectively. Reproduced with permission from reference [11], Copyright © 2015 Elsevier B.V.

It can be seen from Figure 1.7 that a large-scale CO_2 electrolysis plant is not inherently carbon negative. Unfortunately, the concentration of CO_2 in the atmosphere is far too low for applications in an electrolyser, so capture and concentration need to be considered as part of the wider operation [42]. The total process will have significant power requirements before the capture, separation, concentration and electrolysis of CO_2 and also for the subsequent separation of the end product from unreacted CO_2 in the reactor outflow. A number of intuitive steps can be taken in order to lessen the energy requirements in order to push the process towards carbon negative status. Starting with clean and concentrated CO_2 sources will lessen the initial energy requirements. Similarly, a highly efficient reactor will minimise the extraction costs for the product from the outflow gas. This could be through employing cyclic gas flow loops so that unreacted CO_2 is fed back into the reactor, although the recycled gas would have a lower CO_2 concentration, which would result in increased energy requirements for the electrolysis reactor.

Carbon costs associated with energy usage can also be mitigated through hyphenation with green energy sources. Of course, as green energy solutions become more widely employed, the national electrical grid will become increasingly low carbon, which will be beneficial to the overall carbon costs. On a smaller scale, hyphenating CO_2 electrolysers with dedicated solar panels provides a local source of zero-carbon energy. Of course this comes with the usual challenge of energy requirements during off-peak energy production (low sunlight), so some degree of energy supplementation from energy storage solutions or direct from the grid would be required.

A further carbon source that must be considered is in the transport requirements associated with the reactor operation, namely, shipping CO_2 and consumables to the reactor, and then shipping the product to the end user. This highlights the desire to construct reactors onsite for either CO_2 production or product usage in order to remove shipping costs and carbon considerations from the overall operation. This concept of decentralised CO_2 production is receiving growing interest, with the idea that individual green energy providers could hyphenate with an independent CO_2RR unit, either to produce value-added products or biofuel for green energy security [43].

The wider carbon costs of CO_2 electrolysers can also be somewhat offset by considering the presently employed manufacture technique for the target product. CO_2RR products are traditionally sourced from oil via steam reformation. This is a very high-energy process, as well as using fossil fuels as the carbon source. Any viable CO_2 electrolyser will inherently lower global dependency on fossil fuels, which will offset some of the carbon costs highlighted in Figure 1.7.

As was seen for CO_2RR to value-added products, practical applications of CO_2RR for biofuel products must consider the full carbon and energy costs of the overall process. It is also important to consider the energy efficiency for the conversion of CO_2 to biofuel during the storage process and for fuel to energy during later consumption. All real reactors come with some inherent inefficiencies in this conversion. Common losses include low product selectivity, resistive components/IR drop within the reactor and energy lost as heat. Considerations for a CO_2 to biofuel conversion plant are highlighted in Figure 1.8.

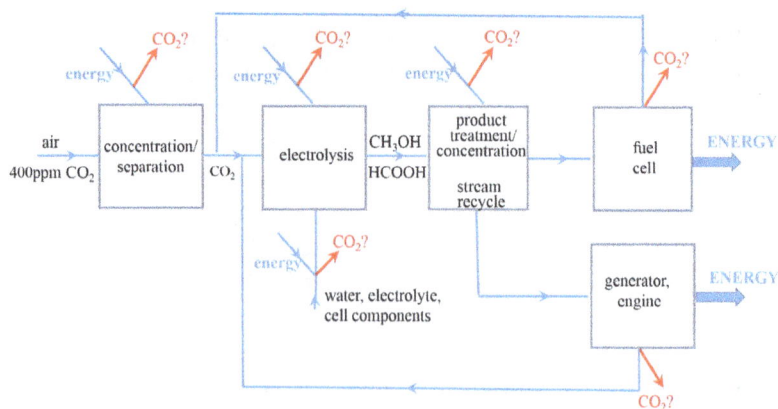

Figure 1.8: Scheme for a plant to remove CO_2 from the atmosphere using a CO_2 electrolyser in order to produce biofuel for energy storage. Energy inputs and potential sources of CO_2 production are highlighted by blue and red arrows, respectively. Reproduced with permission from reference [11], Copyright © 2015 Elsevier B.V.

1.3.4 Financial costs

When considering the role of CO_2RR in the global carbon challenge, it is important to work towards a technology that not only converts CO_2 into a value-added product, but does this in a way that is cost-effective. Using CO_2RR to tackle CO_2 emissions does not only require new technologies, it requires rapid and worldwide uptake of those technologies in place of present carbon-intensive processes. Sadly, simply being more environmentally friendly than fossil fuel–based synthesis routes is not enough to encourage such a large-scale uptake. In order to achieve this, this process must not only be more environmentally friendly than existing routes, but cheaper as well (Table 1.2).

Table 1.2: Market price and annual global production of a number of major CO_2RR products.

Product	No. electrons	Price ($ kg^{-1})	Normalised price $\times 10^3$ ($ $electron^{-1}$)	Annual global production (Mt)
Carbon monoxide (syngas)	2	0.06	0.8	150
Formic acid	2	0.74	16.1	0.6
Methanol	6	0.58	3.1	110
Methane	8	0.18	0.4	250
Ethylene	12	1.3	3.0	140
Ethanol	12	1.0	3.8	77
n-Propanol	18	1.43	4.8	0.2

Adapted with permission from reference [44], Copyright © 2018 American Chemical Society.

Many challenges associated with lowering CO_2RR costs are tackled within the same scope as present research efforts. One of the largest financial considerations is the energy costs associated with CO_2RR, which intuitively should be kept to a minimum. Energy costs vary significantly between different CO_2RR products, with the general trend being that higher order C_{2+} products come with greater energy costs, since there are a larger number of electrons transferred per mole of end product (Figure 1.9).

State-of-the-art catalyst materials reduce the required overpotentials for CO_2RR by modulating the energies of key reaction intermediates bound to the electrode surface. Catalyst and reactor designs can also improve the FE of the system, which increases the proportion of spent energy that goes towards producing the target product, thereby minimising energetic waste.

Of course, a 100% efficient reactor design is highly unlikely for any electrochemical system, so alternative routes are needed to offset the energy costs. Hyphenating the reactor directly into green energy sources such as solar or wind can supplement

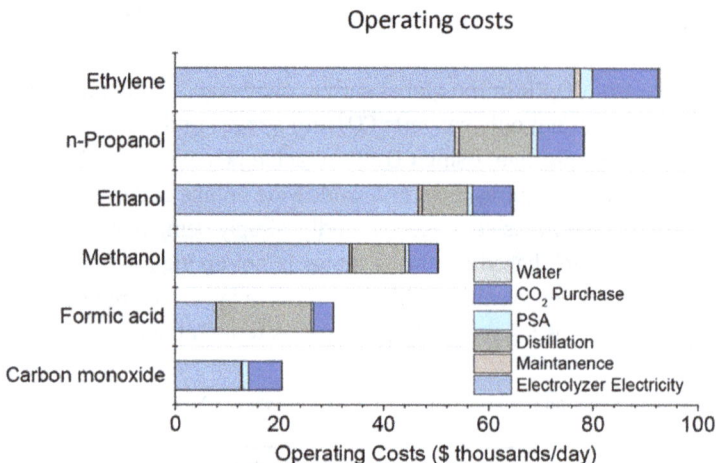

Figure 1.9: Projected operating costs for a CO_2RR reactor producing a number of different possible products, taking a number of best-case scenario assumptions in materials and supply chain costs. Reproduced with permission from reference [44], Copyright © 2018 American Chemical Society.

this cost, although the intermittent power supply from renewably energies means some reliance on grid electrical energy is still to be expected.

A particularly interesting option is moving towards bifunctional electrochemical reactors. Here, CO_2RR at the cathode is balanced by some electrochemical oxidation to a value-added product at the anode. This production of two useful species from the same reactor is a promising route to offsetting the energy costs for more challenging electrochemical systems. This concept is discussed in further detail in Section 6.1.3.

Working with gas-phase mixtures will require costly separation technologies, working either up- or downstream from the CO_2RR reactor or included offsite as part of the supply chain. Although some degree of purification will inevitably be needed for both the gaseous in- and outflow, steps can be taken to reduce the scale of these costs.

Increased CO_2 conversion rates will lower the amount of unreacted CO_2, and increased FEs will lower the concentration of unwanted CO_2RR side products. Sadly, these two factors are almost always working against each other; increased current densities to increase the conversion rate tend to give larger quantities of unwanted alternative CO_2RR products. This is particularly problematic for reactors attempting to produce C_{2+} products, as higher current densities tend to shift the selectivity of CO_2RR reactors towards C_1 products. C_1 reactors still suffer at higher reaction rates themselves due to the parasitic HER.

Separation costs for the CO_2 inflow are largely determined by the CO_2 source. CO_2 from coal or gas power plants contains sizeable amounts of SO_x and NO_x combustion products, which would need to be removed from the flow stream before being fed into the reactor [44]. Current state-of-the-art technologies for CO_2 capture from power plants come with a cost of $70 per ton of CO_2 [45], although improved process design could reduce this to closer to $44 per ton [46]. Alternatively, CO_2 capture could be moved to sites that naturally produce cleaner CO_2 streams, such as fermentation plants [47].

1.3.5 Stability to continued operations

Moving CO_2RR technologies out of the lab and into general usage will require reactors to provide continually high performance over long periods of continuous operation. Acceptable lifetimes can vary from component to component; it is conceivably more acceptable to replace single electrode from an electrochemical reactor than the entire reactor casing, for example. Electrode failure tends to be due to one of four primary causes: catalyst poisoning, catalyst loss, loss of hydrophobicity or electrolyte precipitation (Figure 1.10) [48].

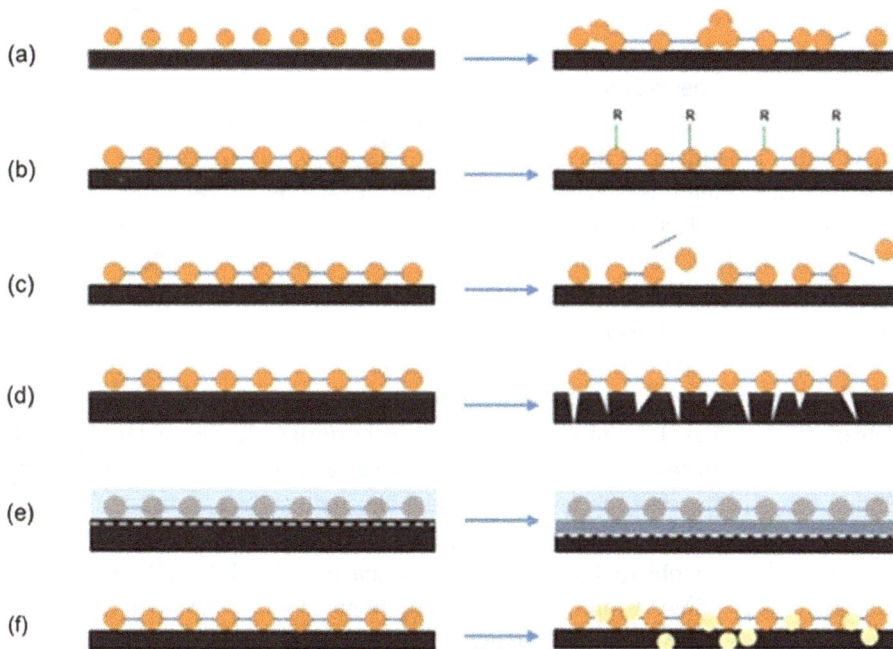

Figure 1.10: Schematic failure modes for catalytic nanoparticles at the surface of a carbon gas diffusion electrode (GDE). (a) Agglomeration of catalyst particles; (b) chemical changes, including catalyst poisoning; (c) binder degradation, potentially leading to dissolution; (d) erosion of the gas diffusion layer by physical or chemical means; (e) loss of GDE hydrophobicity, potentially causing flooding; and (f) carbonate formation and pore blockage. With permission from reference [48], Copyright © 2019 Wiley-VCH Verlag GmbH & Co. KGaA, Weinheim.

Catalyst poisons tend to strongly adsorb metal ions that bind to the catalyst surface, acting to either block CO_2 adsorption sites or change the surface chemistry so that CO_2RR to the desired product becomes disfavoured. These species can be introduced to the reactor even in high-purity electrolytes and can be expected to be more problematic in reactors using CO_2 recovered from industrial sources. In some cases, the reactor can be recovered by chemically or electrochemically treating the electrode to periodically remove such poisons [49]. This is an effective way to recover electrode materials, although the downtime to carry out such recoveries would need to be considered across years of extended operations.

Some degree of catalyst loss over extended operations is difficult to avoid in all electrochemical systems. Material loss by dissolution can normally be avoided by careful electrolyte selection. In the case of metallic catalysts, for example, using a higher pH electrolyte disfavours metal dissolution in favour of oxide formation, following the most thermodynamically favoured phase in their Pourbaix diagram. The mechanical removal of pieces of catalyst material from the electrode surface is commonly seen in flow-by cell configurations, where solution flow can carry off catalytic nanoparticles. Harsh electrochemical conditions such as gas evolution at very negative potentials can also actively dislodge catalytic particles. Similar mechanical wear can also cause degradation to the conductive supports and other electrode and reactor components [48].

Care must be taken to give a strong level of catalyst adhesion and to operate under conditions that can maximise product selectivity, catalyst activity and electrode lifetime. Often this is achieved through incorporating the catalyst with a conductive binder such as Nafion®. Of course, these binders are not impervious to damage themselves, particularly in harsh pH electrolytes. Binder failure can often be seen as a failure in the catalyst at the point where the binder can no longer offer protection from corrosion or delamination [50].

Surface modifications during operation can cause sizeable changes in the reactivity of catalyst surfaces, although catalyst mass is not technically lost. The agglomeration of catalyst particles can occur spontaneously during CO_2RR, which causes a reduced active surface area and a loss in high-energy active sites [51]. Severe agglomeration can also increase the resistivity of the electrode surface, causing increased Ohmic losses over the course of the run [52].

Loss of hydrophobicity has two negative impacts on CO_2RR reactors. The first concerns CO_2 gas flow. A rapid rate of reaction requires a fast supply of CO_2 to the catalyst surface, and also for the removal of gas-phase products, whether desirable (carbon monoxide, methane, ethylene, etc.) or undesirable (non-target gas-phase CO_2RR products and hydrogen) [53]. Gas flow channels are therefore designed to be hydrophobic to prevent them becoming flooded with electrolyte, which would hinder gas transport and give a reduced rate of reaction.

A secondary concern with hydrophobicity is the rate of water electrolysis to give the parasitic HER. CO_2RR specific catalysts are usually immobilised onto conductive supports for integration into CO_2RR reactors. If this conductive support comes into contact with the electrolyte, this will provide a greatly increased electroactive area for the HER to take place. Non-catalytic conductive components should therefore be isolated from aqueous electrolytes, either through hydrophobic treatments or with hydrophobic gaskets. Failure of these measures would lead to a greatly increased HER, which would be seen as a decrease in CO_2RR product selectivity.

The challenge of maintaining electrode hydrophobicity is compounded by the phenomena of electrochemical wetting. This process describes how the surface hydrophobicity of a material at a solid–liquid interface is changed by the application of a potential difference between the solid and the liquid. The process is complex, but briefly, the maximum hydrophobicity of a conductive surface is defined by its point of zero charge. As the potential moves away from this point, the surface becomes less and less hydrophobic [54]. Since CO_2RR requires very negative potentials to operate, some degree to electrowetting is inevitable. Full considerations of the changing hydrophobicity are important, as well as employing engineering controls to protect reactor components from hydrophobicity losses.

Electrolyte precipitation during reactor operation can lead to electrolyte crystals forming either at the catalyst surface or within the structure of 3D porous electrode materials [55]. Surface-bound precipitates can block catalyst active sites to slow the rate of CO_2RR, and precipitates within flow channels can hinder gas flow to and from the electrode surface. Most electrolytes are employed well below their solubility limits, but precipitation is still problematic due to the highly dynamic reaction environment at the electrode surface [40]. Transient changes in solution composition within the diffusion layer, particularly due to local pH changes, cause significant changes to the electrolyte composition, as discussed in Section 1.3.2. CO_2RR reactors much consider not only the starting conditions but the evolution of the electrode–electrolyte interface in order to prevent losses due to electrolyte precipitation.

1.3.6 Key targets

Advances in catalyst, electrode and reactor design are continually working towards improved product selectivity at higher current density. These two parameters are essential to increase in tandem; large current densities mean that production rates are large enough to produce meaningful amounts of product, while high selectivity lowers energy costs for the electrolysis and financial costs associated with separating one target product from a mixture in the outflow. This is unfortunately challenging, as increasing the current density inherently lowers the selectivity in most CO_2RR systems, since larger current densities result in an increased rate of water electrolysis to produce H_2. This is particularly problematic for higher order products such as methane and ethylene as, without targeted modifications to the reactor design, higher current densities also favour the release of lower order CO as the dominant product [36].

At this point, it is useful to put target values for the current density and product selectivity in place as metrics to aim for before wider applications and industrial up-scaling are possible. A number of technoeconomic models have been created, which take into account the difference between costs in constructing and operating a CO_2RR reactor versus potential profits for a range of CO_2RR products [44, 56–62]. Models such as these consider costs across the whole lifetime of the CO_2RR reactor, so stability to continued operations becomes an additional parameter that must be considered; a high selectivity and active catalyst material is not suitable if electrode surfaces have to be replaced every hour. It is also important to consider the overpotential required to drive CO_2RR, as this will impact the energy costs of the final reactor.

As a broad definition, a system could be described as economically viable if it can operate at ~ 300 mA cm^{-2} with a selectivity of ~ 70% and an overpotential of ~ 0.5 V [44]. It is worth mentioning two caveats to this definition. First, the operating costs to viable large-scale production will be different for different products, since these will require different catalyst materials and energy inputs as well as different market prices. Second, the excellent performance in one of these catalysts could off-set flawed performance in others; a low current density material could still be useful on an industrial scale if it shows the same performance for >10,000 h (Figure 1.11).

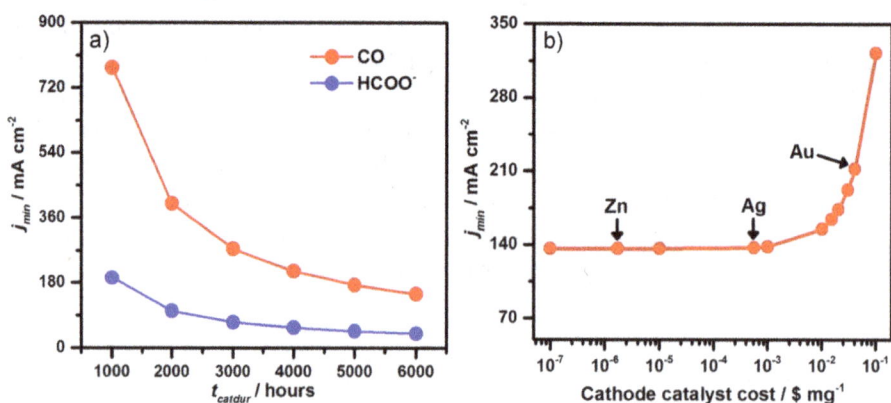

Figure 1.11: (a) Minimum operating current densities (j_{min}) for the production of CO and formate versus the durability of the catalysts (t_{catdur}). Shorter lifetimes require larger operating current densities to compensate. Calculated vales assume a 100% selective reaction at the cathode. (b) Minimum operating current densities (j_{min}) for the production of CO versus the cost of the cathode catalyst. Higher catalyst cost must be offset by a larger current density. Calculated values assume a 100% selective reaction at the cathode with a 4,000 h lifetime. Adapted with permission from reference [61], Copyright © 2016 Wiley-VCH Verlag GmbH & Co. KGaA, Weinheim.

2 Reaction mechanisms

The first step of CO_2RR is the single electron transfer to give the CO_2^- radical anion [41]:

$$CO_2 + e^- \rightleftharpoons CO_2^{\bullet-}, \qquad E^0 = -1.85 \text{ V versus RHE} \qquad (2.1)$$

The value for E^0 is highly dependent on the reaction environment, particularly the solvation of the anion radical and its impact on the radical stability; values of −2.21 and −2.14 V versus saturated calomel electrode have been reported in dimethylformamide and water, respectively [63]. A basic overview of the dominant CO_2RR pathways is given in Figure 2.1.

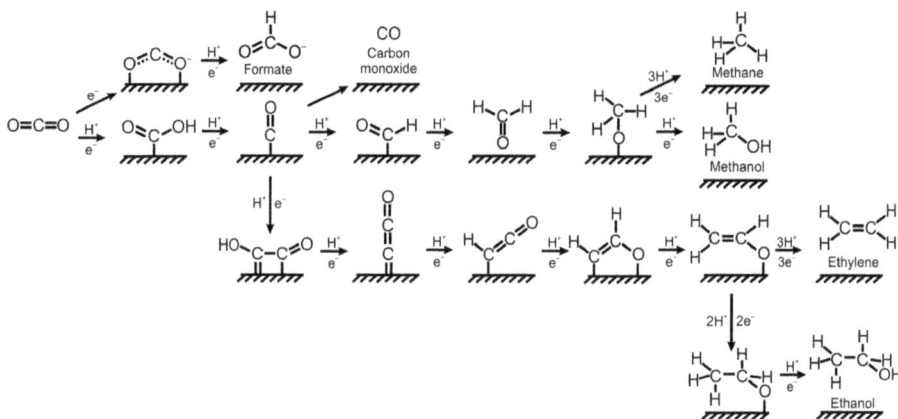

Figure 2.1: Schematic representation of the possible reaction mechanisms for CO_2RR to formate, carbon monoxide, methane, methanol, ethylene and ethanol based on the lowest energy pathways from computational studies [64]. For simplicity, differentiations are not made between proton-coupled electron transfer and electron transfer with a subsequent protonation step. Reproduced with permission from reference [36], available open access under the Creative Commons CC BY license, Copyright © 2020 Elsevier B.V.

2.1 C_1 product route

The nature of the adsorbed CO_2^- radical anion and how it then proceeds to react plays a major role in defining which CO_2RR product will be generated. The electrochemical path to C_1 products is initially defined by the nature of the bond between the catalyst surface and the CO_2^- radical anion. In fact, the strength of the metal–CO_2^- bond is an excellent predictor for whether the reaction will produce two of the most common C_1 products: formate or carbon monoxide.

https://doi.org/10.1515/9781501522239-002

2.1.1 Formate and formic acid

The first situation to consider is a weak bonding, where there is little or no interaction between CO_2^- and the electrode surface. This condition favours the production of formate as the dominant CO_2RR product. To date, three different mechanisms have been proposed for CO_2RR to formate, the key difference being the identification of key adsorbed surface intermediates [12]. The first proposal begins with a proton-coupled electron transfer to CO_2 to give an adsorbed *COOH species [65]. The second begins again with the proton-coupled electron transfer, but this time adsorption is via oxygen to give an *OCOH intermediate [66]. For both of these reactions, the adsorbed species undergoes a second proton-coupled electron transfer to liberate the formate end product.

The final process involves an initial electron transfer to a proton itself to give *H. CO_2 then reacts with *H via a proton-coupled electron transfer in the following step in order to produce formate [67]. The *H pathway has been supported by experimental observations, where adsorbed carbonate was not detected at bismuth electrodes. However, computational evidence suggests that the *OCOH pathway is the most energetically favourable (Figure 2.2), so this will be the pathway further discussed here. These are detailed as follows, where * indicates a surface site taking part in the reaction:

$$* + CO_2 + H^+ + e^- \rightleftharpoons \text{*OCOH} \qquad (2.2)$$

$$\text{*OCOH} + H^+ + e^- \rightleftharpoons \text{*HCOOH} \qquad (2.3)$$

$$\text{*HCOOH} \rightleftharpoons HCOOH + * \qquad (2.4)$$

Formic acid cannot be reduced to any alternative CO_2RR products. This indicates that the CO_2RR mechanism to formate exists along an entirely separate pathway to the other possible products. The origin of this singular reduction route is the unique binding mode between CO_2 and the catalyst surface for formate production versus all other CO_2RR products. In all cases, CO_2 adsorbs in a bent configuration with a symmetric orientation with respect to the catalyst surface [68]. For other hydrocarbon products, CO_2 adsorbs in a carbon-down orientation, whereas the formate route begins with CO_2 adsorbed in an oxygen-down orientation [12].

The binding energy of *OCOH is therefore a key factor in determining the activity of a catalyst material towards formate over other CO_2RR products. Feaster et al. demonstrated this by plotting the partial current density for CO_2RR towards formate against the *OCOH binding energy, which shows a clear volcano-type relationship with tin electrodes closest to the ideal binding energy (Figure 2.3).

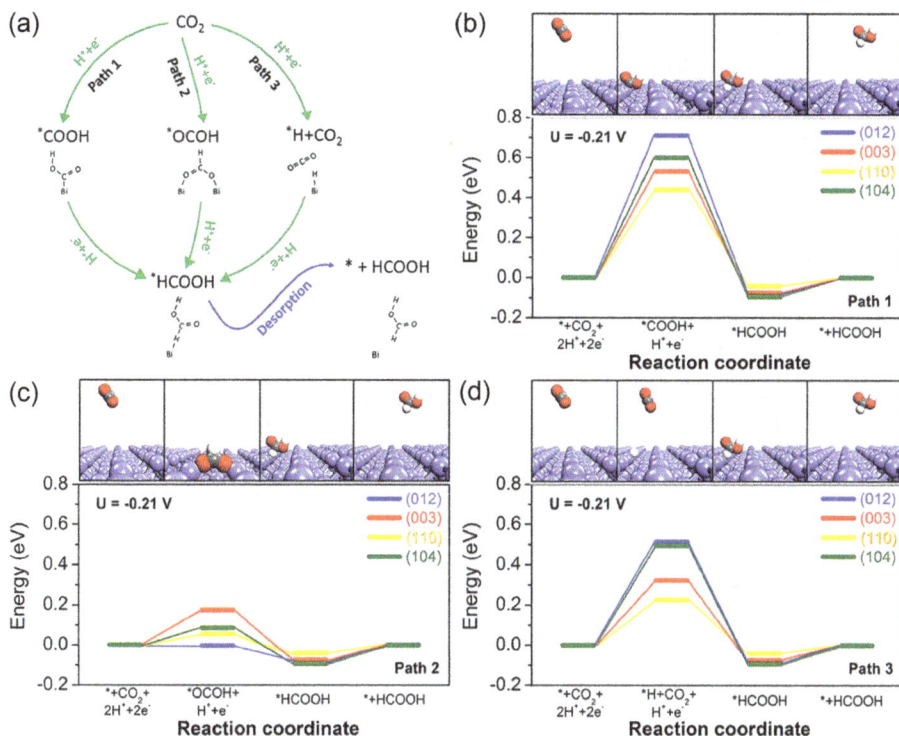

Figure 2.2: Schematic of three possible mechanisms for the formation of formate via CO_2RR at bismuth electrodes. Density functional theory calculations by the authors suggested that the oxygen-bound pathway via *OCOH is the lowest energy, and therefore the most likely to proceed. Adapted with permission from reference [69], Copyright © 2017 American Chemical Society.

2.1.2 Carbon monoxide

Stronger interactions between the catalyst and the CO_2^- radical anion favour the CO product pathway. Recent computational works from the Nørskov group suggest that, for CO_2RR at transition metal electrodes, the first electron transfer to CO_2 occurs as a proton-coupled electron transfer, that is, CO_2 is reduced and protonated in one single step to give *COOH as the first reaction intermediate [71]. *COOH is then further reduced via a single electron transfer to give *CO at the electrode surface:

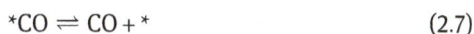

$$* + CO_2 + H^+ + e^- \rightleftharpoons *COOH \tag{2.5}$$

$$*COOH + H^+ + e^- \rightleftharpoons *CO + H_2O \tag{2.6}$$

$$*CO \rightleftharpoons CO + * \tag{2.7}$$

From a mechanistic point of view, the selectivity of CO_2RR toward CO is strongly determined by the rate of reaction (2.7). Ideally, this reaction should proceed quickly,

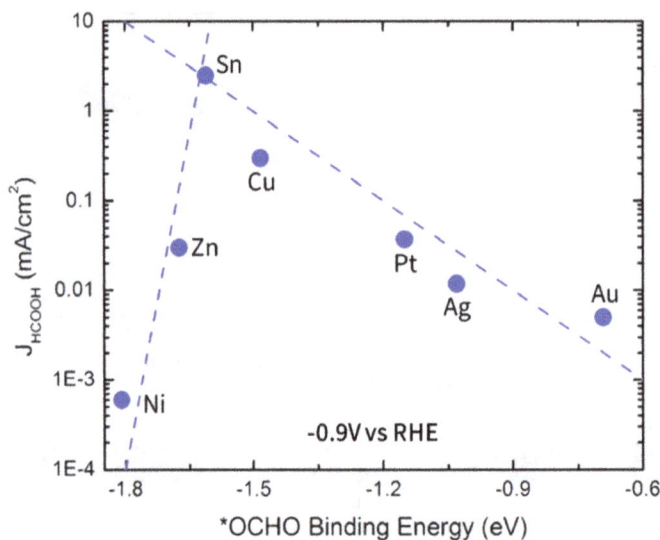

Figure 2.3: Volcano plot showing the relationship between *OCHO binding energy and the partial current density for formate production by CO_2RR (j_{HCOOH}). Note the log scale indicating the sizeable impact of *OCOH binding energy on the rate of CO production. Reproduced with permission from reference [70], Copyright © 2017 American Chemical Society.

so that CO detaches from the electrode surface as soon as it is formed. This has a dual benefit since (i) CO is removed from the catalyst surface before any further electron transfers can occur and (ii) catalytic sites (*) are rapidly freed up for CO_2 to adsorb, increasing the catalyst turnover rate. However, the best computational descriptor for CO selectivity is the adsorbed *COOH species, since this also incorporates selectivity for CO versus formate (Figure 2.4).

Unlike formate, CO can be further reduced to alternative CO_2RR products. In fact, focussing on CO reduction is a viable means to better understand the CO_2 reduction mechanism, where the mechanism can be probed assuming that the first two electrons have been transferred with high efficiency [72]. The following sections focus on higher order CO_2RR products, and so mechanistic steps following from the generation of CO will be a common theme.

2.1.3 Methane and methanol

The starting point in the CO_2RR mechanism to methane is the formation of CO*, which remains at the electrode surface rather than diffusing away to be collected as the product. Computational approaches have then probed the relative thermodynamic and kinetic barriers for the following possible intermediates in order to predict the most likely mechanism. Based on these, different groups have suggested

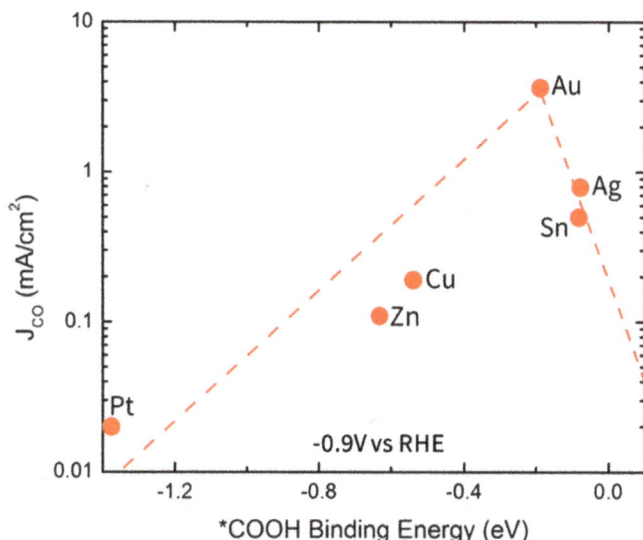

Figure 2.4: Volcano plot showing the relationship between *COOH binding energy and the partial current density for CO production by CO_2RR (j_{CO}). Note the log scale indicating the sizeable impact of *COOH binding energy on the rate of CO production. Reproduced with permission from reference [70], Copyright © 2017 American Chemical Society.

multiple possible reaction routes across a range of different intermediates. The different results usually stem from the different assumptions made in the construction of the varied computational models employed to probe the CO_2RR mechanism. There are a number of excellent reviews and articles available that go into detail regarding the complexities of the different possible models and the impact of their design on the computational outputs [64]. In these cases, differentiation between theoretical routes is facilitated by extrapolating theoretical predictions to experimental observations.

For example, computational studies by Lui et al. have provided insights into the precise route and rate-determining step for CO_2 reduction to CH_4 using density functional theory (DFT) [73]. The authors constructed a free energy diagram, detailing the energy change across a series of elementary steps, starting with CO_2 and ending with CH_4. The potential dependence of the reaction can clearly be seen when comparing the high-energy route at 0 V versus RHE (reversible hydrogen electrode) to the low-energy route at −0.5 V versus RHE (Figure 2.5, black and red, respectively).

The authors also demonstrate two different routes for moving between key intermediates *CO and *CHOH. The hydrogenation of *CO to give the alcohol *COH has a much higher activation energy barrier than its hydrogenation to the aldehyde *CHO, indicating that the reaction mostly likely proceeds via the latter intermediate. In either case, this hydrogenation step can be seen to have the largest activation energy barrier across the entire mechanism, indicating that this is the rate-determining step.

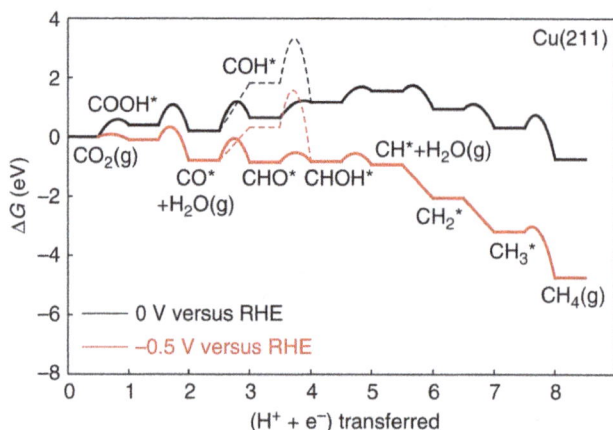

Figure 2.5: Mechanistic steps for the reduction of CO_2 to methane on a stepped Cu(211) surface via a number of adsorbed intermediates. ΔG indicates the energy change moving from one species to the next. Reading along from left to right shows the activation energy required for each step. The power of density functional theory in defining reaction mechanisms is highlighted by comparing the solid versus dashed routes between CO* and CHOH*. The solid line assumes the reaction follows CO* → CHO* → CHOH*, whereas the dashed assumes CO* → COH* → CHOH*. The much lower energy barrier for the solid route suggests that this is the correct mechanistic route. Adapted with permission from reference [73], available open access under the Creative Commons CC BY license, Copyright © 2017 X. Liu et al.

It is worth mentioning at this point that such studies are surface specific; Figure 2.5 suggests that the hydrogenation of *CO to *CHO is the rate-determining step specifically for the Cu(211) electrode surface, and calculations from Peterson et al. and Nie et al. revealed very different mechanisms for the (211) and (111) faces.

DFT calculations by Peterson et al. at the same Cu(211) surface suggested that it is more thermodynamically favourable to maintain the second C–O bond for as long as possible throughout the CO_2RR to CH_4. The sequential reduction and hydrogenation of *CO leads to the formation of the *H_3CO methoxy intermediate. *H_3CO is then reduced to liberate CH_4 leaving a surface-bound *O, which undergoes a two-electron reduction to H_2O to free up the catalytic site [74]. Alternative calculations from Nie et al. predicted the mechanism at Cu(111) proceeds via a *C intermediate formed by the removal of both O atoms of CO_2 to give $2H_2O$. *C then undergoes four proton and electron transfers to give CH_4 [75].

Moving onto other CH_4 producing materials show similar mechanistic routes with CO_2 first reduced to *CO, then to *CHO or *COH, which is then reduced to either CH_4 or CH_3OH [76]. The primary difference between Ag and Cu in this case is that CO bonds with Ag more weakly than Cu, so Ag would likely give CO as the major product despite having an active CH_4 route.

Within this C_1 route, the final consideration is for the selectivity between CH_4 and CH_3OH. Different proposed mechanisms convert CO_2 to CH_4 via very different

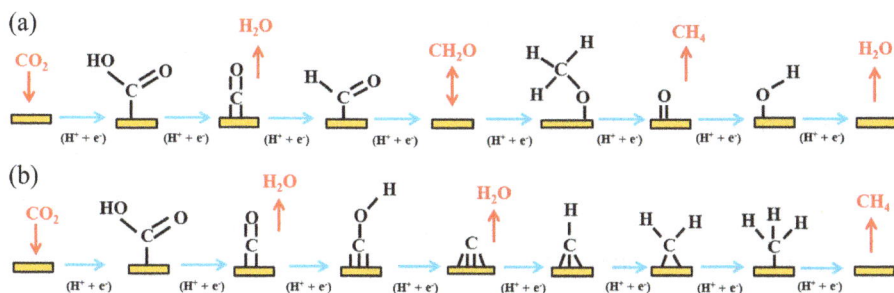

Figure 2.6: Pathways for the electrochemical production of methane from CO$_2$ on Cu electrodes from (a) a thermodynamic analysis at Cu(211) (adapted from reference [74]), and (b) a combined thermodynamic and kinetic analysis from reference [75]. Species in black are adsorbates, while those in red are reactants or products in solution. Reproduced with permission from reference [64], Copyright © 2015 American Chemical Society.

intermediates, so clearly the different mechanisms will offer different key steps that define the selectivity between CH$_4$ and CH$_3$OH. Following Peterson et al.'s proposed mechanism (Figure 2.6a), the adsorbed *CH$_3$O species is a logical diverging point for CH$_4$ versus CH$_3$OH, depending on the *–O or C–O bond that is cleaved as a result of the electron transfer. Alternative studies by Schouten et al. suggested that there is a direct reduction path from *CH$_2$O to methanol that is in competition with the methane producing mechanism [77]. In either case, fine-tuning the bonding be-tween catalyst and the adsorbed intermediate is key for selectively producing either CH$_4$ or CH$_3$OH over the other.

2.2 C$_2$ product route

2.2.1 Ethylene and ethanol

As with the mechanism for CH$_4$ formation, mechanistic studies of C$_2$ product are dominated by copper surface, since copper is unique in producing large quantities of C$_2$ products during CO$_2$RR. It is worth mentioning that C$_2$ products can be pro-duced at other materials, such as Ni$_3$Al and PdAu. However, none to date are any-where near the selectivity of even very simple Cu materials, so it is not surprising that computational efforts have focused on Cu surfaces [36].

The key intermediate for C$_2$ product formation is the adsorbed *CO species. A weak catalyst–CO bond allows CO release as the majority product, whereas a stron-ger catalyst–CO bond holds CO for further electron transfers. Computational and ex-perimental CO reduction studies can therefore provide insights into CO$_2$ reduction mechanisms and properties [78]. From this point, the selectivity for C$_1$ versus C$_2$ is defined by the likelihood of C–C bond formation during those electron transfer steps. Different mechanisms have been proposed for how this bond formation occurs. One commonality in all studies is the presence of *CO as part of the C_2 mechanism, which has been detected spectroscopically [79]. However, the mechanism of C–C bond formation is not straightforward.

The two most commonly described mechanisms are the low overpotential pathway, most commonly seen at the Cu(100) facet, and the high overpotential pathway, most commonly seen at the Cu(111) facet [80]. The low overpotential route proceeds via a dimerisation step between *CO species as the rate-determining step [81]. This is supported by spectroscopic detection of *CO–COH, which is the expected result of that dimerisation with subsequent proton and electron transfer [79].

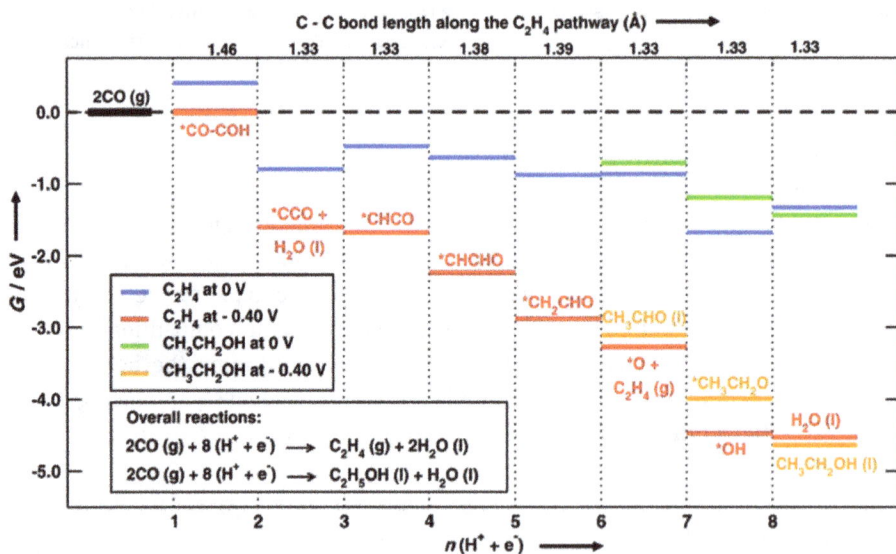

Figure 2.7: Lowest energy pathways for the electrochemical reduction of CO to C_2H_4 and C_2H_5OH at 0 and −0.4 V versus CHE. Reproduced with permission from reference [81], Copyright © 2013 WILEY-VCH Verlag GmbH & Co. KGaA, Weinheim.

Figure 2.7 offers an explanation for the observed trend at Cu electrodes that C_2H_4 is usually produced in larger quantities than C_2H_5OH, since the sixth electron transfer is more energetically favoured for the C_2H_4 pathway (red) than for the C_2H_5OH pathway (yellow) by around 0.2 eV. Of course, this is only a thermodynamic perspective but the agreement with experimental observations is convincing. Within this mechanism, the authors highlight that the dimerisation between *CO and CO occurs with a separated proton and electron transfer. This is distinct from proton-coupled electron transfers that are seen in the CO_2 to CH_4 mechanism. This is important, since experimental observations at Cu(100) surfaces show that the CO_2RR mechanism to CH_4 is pH dependent, whereas the mechanism to C_2H_4 is independent of pH. This presents experimentation at high pH as a viable means to favour C$_2$ product formation over the production of significant amounts of CH$_4$ [82].

This mechanism is defined as the low-overpotential route due to the formation of the CO–CO dimer being more energetically favourable at Cu(100) surfaces [83]. The rate of dimerisation is also accelerated by a greater surface coverage of *CO on Cu(100) at low overpotentials. At higher overpotentials, CO dimerisation becomes increasingly disfavoured, until coupling between *CO and *CHO becomes the new lowest energy pathway [84]. The same trend is seen when moving from Cu(100) to Cu(111) facets, with Cu(111) giving significantly higher energy barrier to the CO dimerisation, leading to a more favourable mechanism via *CHO (Figure 2.8) [85].

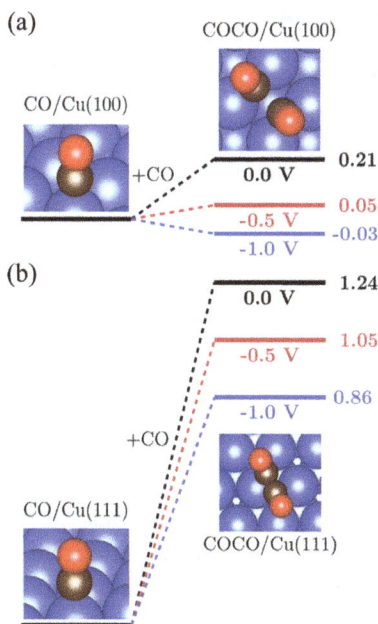

Figure 2.8: Free energy diagrams for CO dimerisation on (a) Cu(100) and (b) Cu(111) surfaces at three potentials (*E* vs RHE at pH = 7). Energy levels given in eV. Reproduced with permission from reference [85], Copyright © 2018 American Chemical Society.

The most noteworthy difference between the high and low overpotential mechanisms is the presence of *CHO species. Importantly, *CHO is also a key intermediate for CO$_2$RR to CH$_4$. This explains the observation that C$_2$H$_4$ can be produced at low overpotentials without any CH$_4$ generation, whereas higher overpotential production, or production at Cu(111) facets, gives simultaneous CH$_4$ and C$_2$H$_4$ production [86]. This is problematic for up-scaling C$_2$H$_4$ production, since an industrial reactor would need to run at high overpotentials in order to provide useful quantities of C$_2$H$_4$. High overpotential C$_2$H$_4$ production must therefore involve as much consideration to CH$_4$ hindrance as to C$_2$H$_4$ production in order to selectively produce the target product.

This difference also needs to be taken into account when considering different types of copper materials. Polycrystalline copper contains mostly (100) facets, which would therefore favour C_2H_4 production. However, the same is not necessarily true for micro- or nanostructured copper materials, which can be engineered to a specific crystal facet [80]. In the selectivity challenge of C_2H_4 versus C_2H_5OH, Figure 2.7 shows that the mechanism is identical up until the sixth electron transfer to *CO. The selectivity for C_2H_4 versus C_2H_5OH is therefore determined by whether the adsorbed *CH_2CHO undergoes a hydrogenation to continue the alcohol mechanism, or a dehydroxylation to release C_2H_4 as the final product.

On the other hand, computational studies by Goddard et al. proposed that the selectivity is defined by much earlier intermediates. They identified reaction steps to the *CHCHO intermediate for determining C_2H_4 versus C_2H_5OH selectivity [87, 88]:

$$*CHCHO + H^* \rightleftharpoons *CHCHOH \tag{2.8}$$

$$*CHCHO + e^- \rightleftharpoons *CCH + OH^- \tag{2.9}$$

This presents a means of producing alcohol-selective materials by designing catalyst surfaces to favour the production of reactive *H to drive the hydrogenation of *CHCHO intermediate [78].

2.2.2 C_{3+} and beyond

Electrochemical production of C_{3+} carbon products is far more challenging than for C_1 or C_2 materials. From a mechanistic perspective this is intuitive, since C_{3+} products must pass through a significantly larger number of adsorbed intermediates than for C_1 or C_2. Many of these intermediates will be shared with mechanistic routes to species that require fewer electron transfers, so C_{3+} products see significantly lower production rates than competing C_1 and C_2 alternatives. As such, computational studies into C_{3+} production mechanisms are considerably fewer and further between.

Discussions into the C_{3+} production mechanism focus on opportunities for further C–C bond formation. These usually involve the insertion of a relatively low-order CO_2RR intermediate such as *CO or *CHO into higher order intermediates such as *CHCHO or *CH_2CHO [89].

A certain degree of mechanistic information can be inferred from the nature of C_{3+} products being formed. Kuhl et al. detected hydroxyacetone (CH_3COCH_2OH) for CO_2RR at a copper electrode, which is the C_{3+} product that requires fewest proton and electron transfers. This suggests that coupling between C_1 species early in the mechanism is feasible [90]. Kortlever et al. detected CO_2RR products up to C_5 at PdAu catalysts. They proposed that this was due to the polymerisation of surface-bound *CH_2 species made possible, thanks to the tuned CO binding energy at their material surface [91].

Although the exact mechanism is still unclear, finding innovative ways of facilitating the coupling of CO_2RR intermediates is vital in achieving C_{3+} production.

2.3 Computational activity predictors and scaling relationships

One of the key motivators for the wealth of computational CO_2RR studies is the search for CO_2RR catalytic activity descriptors. These descriptors are relatively simple parameters that are proportional to the activity of a certain material or surface towards CO_2RR. If these descriptors can be well defined and understood, this presents an opportunity to parametrically refine, modify and discover new CO_2RR materials using a computational approach. This presents a significant saving in both time and cost, as experimental studies could focus their attention on a handful of the most promising candidates as defined by the computational descriptor, rather than having to synthesise, assemble and analyse each material in turn.

For reversible electrochemical systems, this is a relatively straightforward process. Considering the heterogeneous redox of some species A to give B,

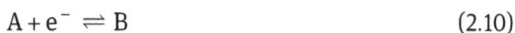

$$A + e^- \rightleftharpoons B \qquad (2.10)$$

it is highly likely that catalysts exist that can carry out both directions of this reaction (i.e. the oxidation and reduction) with minimal overpotential [92]. Bearing in mind that we have defined this as an electrochemical reaction to a surface-bound species, we could rewrite eq. (2.10) in terms of various surface intermediates:

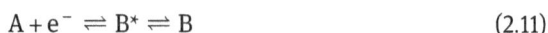

$$A + e^- \rightleftharpoons B^* \rightleftharpoons B \qquad (2.11)$$

where * indicates that the species is bound to the electrode surface. In order for this reaction to proceed efficiently at an electrode surface, the strength of the bonding to A and B be of an intermediate, compromise value. If the bonding is too large, then the electrode surface can become poisoned by B*, that is, the surface will be completely covered by B*, leaving no free surface sites for A to be reduced. If the bonding is too weak, then adsorption of A will not be favoured, giving a low availability of A for reduction and therefore a slow rate of reaction.

This presents the opportunity to predict the activity of a material towards a specific electrochemical reaction by plotting the activation energy of the rate-determining step against the binding energy of the intermediate species to the catalyst surface. The resultant volcano plot will then give the optimal materials at the peak of the volcano at the desired intermediate binding energy.

While relatively straightforward for a simple reversible system, this approach becomes significantly more complicated when moving towards a complex, multistep electrochemical reaction. Reactions involving multiple electron transfers, such as CO_2RR, proceed via multiple surface-bound intermediates. This requires researchers to dig deeper into the reaction mechanism in order to look for scaling relationships between key intermediates and transition states in order to find one key species that may be used as a descriptor for the entire reaction. This trend of needing an intermediate binding energy is seen not only for CO_2RR materials but also for oxygen reduction, water oxidation and indeed any surface-dependent electrochemical process [93].

As we have seen in the previous sections of this chapter, CO is a key intermediate in CO_2RR. Subsequent steps after the adsorbed *CO is formed play a defining role in the selectivity of a material towards a specific CO_2RR product. Computational studies have also looked into *CO as the starting point for the rate-determining step for CO_2RR to higher order products [73]. It is therefore not surprising that *CO has been identified as a key CO_2RR descriptor. Liu et al. showed a clear linear correlation between free energy of the CO_2RR transition state (H–CO*) and CO adsorption energy for CO_2RR to methane at a number of different materials [73].

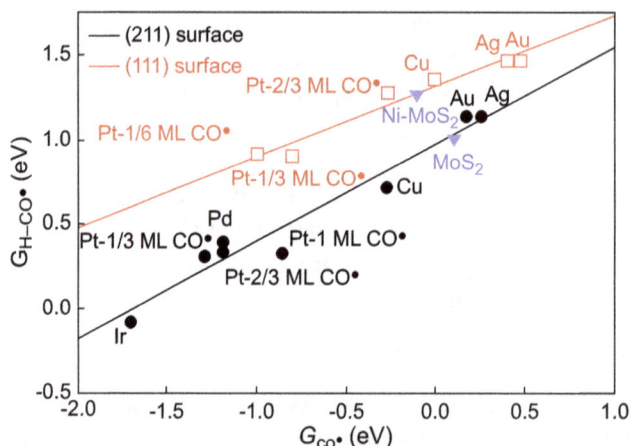

Figure 2.9: Scaling relationship between the CO reduction transition state energy (G_{H-CO*}) and the binding energy for CO (G_{CO*}) for the (111) (red) and (211) (black) facets of a number of metals. A good linear relationship is seen for both surfaces. Adapted with permission from reference [73], available under Creative Commons CC BY license Copyright © 2017 Liu et al.

Since the rate-determining step starts with adsorbed CO* and passes through the transition state H–CO*, the free energies of these two species can be used to express the activation energy for the reaction (E_a):

$$E_a = E_{H-CO^*} - E_{CO^*} \tag{2.12}$$

Based on this relationship, the authors were able to show a volcano-type relationship for both the activity of a material towards CO reduction and the selectivity of the material towards CO reduction versus hydrogen evolution. Note that the authors here discuss materials in terms of CO reduction rather than CO_2 reduction, since

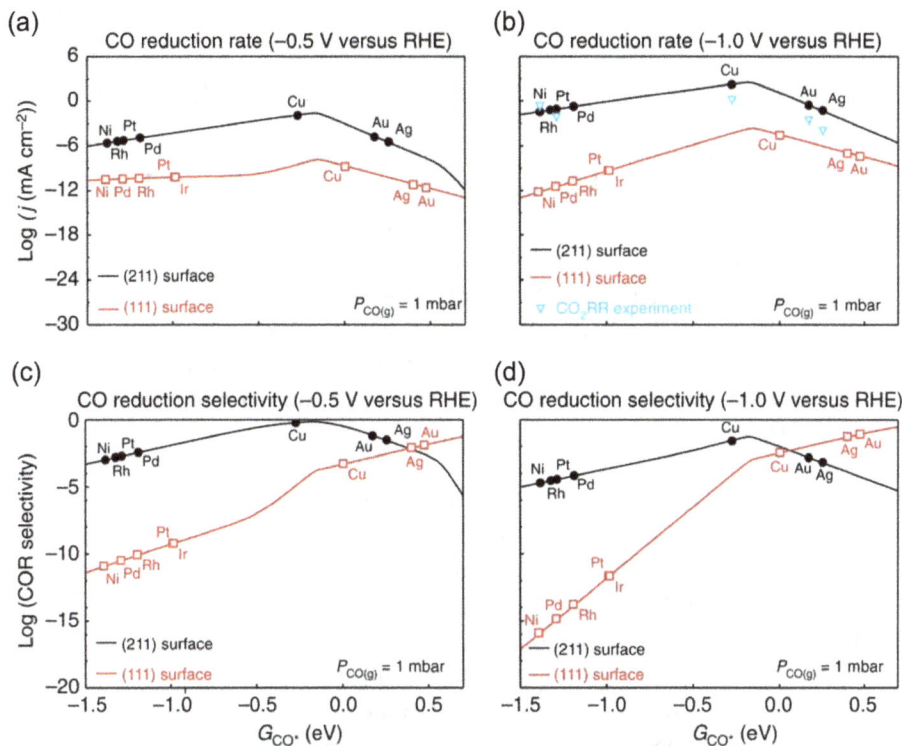

Figure 2.10: Activity and selectivity volcanoes for CO reduction at (211) (black) and (111) (red) metal surfaces. Top row shows the CO reduction activity as the log of the current density (j). Bottom row shows the selectivity towards CO reduction versus H_2 evolution, where COR selectivity is defined as the ratio of the CO reduction current versus the total current passed. The left and right columns are simulated at −0.5 and −1.0 V versus RHE. Blue points are select experimental data, which align well with the computational results. Adapted with permission from reference [73], available under Creative Commons CC BY license Copyright © 2017 Liu et al.

they had defined the rate-determining step as the hydrogenation of CO at the electrode surface under these particular conditions.

From Figure 2.10, we can infer a number of key factors regarding the CO_2RR mechanism:

(i) The CO_2RR activity is substantially higher at stepped (211) facets compared to flat (111) facets. This is consistent with experimental observations that reaction rates at electrochemical catalysts are much faster at highly defective surfaces than smooth ones.

(ii) The high activity of the stepped (211) facet can be related to the lower activation energies shown in Figure 2.9. The authors suggested that this is due to the increased accessibility of the carbon to an incoming proton and the ease of *CO rotation towards the transition state lowering G_{H-CO*}.

(iii) Copper sits at the peak of the volcano for both activity and selectivity. Materials that bind CO more weakly than Cu are limited by low CO* availability, giving a steep drop in activity moving towards Ag and Au. Materials that bind CO more strongly than Cu risk becomes poison by CO* or other intermediates. The binding energy is so strong that absorbed carbonaceous materials are not released giving a low catalyst turnover rate. The difference in gradient of the two slopes to the volcano predicts that low CO* coverage has a more detrimental effect than the risk of surface poisoning.

(iv) Only Cu(211) shows a viable selectivity towards CO reduction over the competing H_2 evolution. Increasing the potential from -0.5 to -1.0 V versus RHE increases the activity of the materials but lowers the selectivity. This supports the experimental observations that high rates of reaction inevitably result in large concentrations of H_2 gas.

The position of the peak of the volcano is at a greater overpotential than the thermodynamic equilibrium potential for the reaction. The origin of this overpotential comes from scaling relationships between the energies of two key intermediates due to similarities between their bonds [64]. In the case of CO_2RR to methane this is CO* and CO–H* [94]. Looking at eq. (2.12), one can imagine an approach to reducing the activation energy of the reaction (E_a) by simply lowering the binding energy of the H–CO transition state ($E_{H–CO}$). However, the linear scaling relationship between $E_{H–CO*}$ and E_{CO*} means that changes to $E_{H–CO*}$ will cause a corresponding shift in E_{CO*}, resulting in an unchanged E_a.

The situation becomes more complex when considering the full CO_2RR reaction mechanism to all of the possible products. The complete mechanism passes through a large number of possible intermediates (see Figure 2.1). This creates a large number of scaling relationships that must be considered, depending on the target product: *OH scales with *OCH_3 in the route to methane and methanol, *OH scales with *$OCHCH_2$ in the route to ethylene and *$OCHCH_2$ scales with *OCH_2CH_3 in the route to ethanol (Figure 2.11) [74, 75, 81].

A full mechanistic approach to the design of new catalyst materials for CO_2RR must therefore design a specialised catalyst surface that is able to break these scaling relationships in order to reach higher order rates of reaction [71, 94, 95]. A number of different design strategies have been proposed to meet this target. One option is to design a metal alloy catalyst so that different stages of CO_2RR occur at different catalytic sites, that is, on different metal atoms within the alloy. This so-called tandem catalysis has been designed so that the initial two-electron CO_2RR occurs on "site A" before CO is passed onto "site B" for further reduction to a higher order CO_2RR product such as methane or ethylene [96]. This effectively removes the scaling relationship between *CO and *CHO because the intermediates are now important at different metal sites; *CO can be optimised at site A while *CHO is optimised at site B.

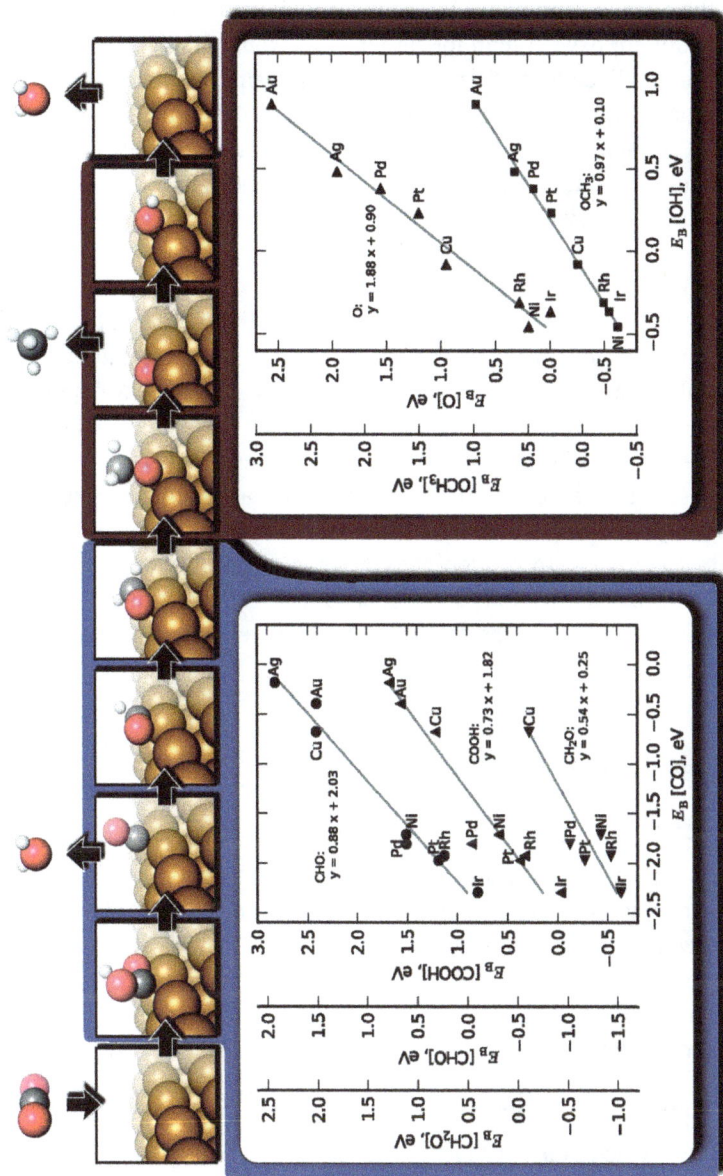

Figure 2.11: Scaling relationships for C-bound intermediates *CHO, *COOH and *CH$_2$O relevant to the CO$_2$RR pathway on a number of different metal catalyst surfaces. Reproduced with permission from reference [94], Copyright © 2012 American Chemical Society.

Alternately, the alloy catalyst could be produced with a second element with a higher binding affinity for oxygen than for carbon. This would allow *CHO to bind to the catalyst material via both the C and O atoms, increasing its stability without negatively impacting the binding energy of *CO [97]. This break in the scaling relationship between *CHO and *CO would give a decreased activation energy for CO_2RR to methane. However, other impacts must still be considered here. Increasing O binding affinity may increase the overall CO_2RR potential by promoting catalyst poisoning by *OH [94].

The same core concept of stabilising O can be achieved via surface-bound tethered ligands. These would preferentially stabilise the *CHO through ligand–oxygen interactions, while having no impact on the stability of *CO. This is a technique that has been successfully employed for homogeneous catalysis and may have applications for heterogeneous catalyst as well [98]. Here, care must be taken to make sure that there is sufficient surface-bound ligand to give the desired stabilisation without blocking too many surface sites to limit the overall electrocatalytically active surface area. The adsorption properties may be more directly altered through the use of promoters at the catalyst surface. Adsorption onto the promotor will have distinct energy characteristics versus adsorption to the bulk catalyst. This can help to break the scaling relationship, so long as the promoter chosen is resistant to the strongly reducing CO_2RR conditions.

2.4 Hydrogen evolution reaction

The reduction of water to give hydrogen gas via HER is a common occurrence in all aqueous CO_2RR systems. In many cases, a fair improvement in the CO_2RR rate and in the product selectivity that can be achieved by designing a system that can hinder the HER, even without making considerations for the CO_2RR itself. It is therefore useful to understand the mechanism of the HER that proceeds at a catalyst surface. The overall reaction itself is straightforward and can be described as a two-electron reduction of two protons [99]:

$$2H^+ + 2e^- \rightleftharpoons H_2, \quad E^0 = 0.00V \text{ versus RHE} \tag{2.13}$$

This reaction can be divided into three elementary steps. The first is the Volmer reaction, where a single proton is reduced to give the adsorbed *H at the electrode surface:

$$H^+ + e^- \rightleftharpoons {}^*H \quad \text{(Volmer reaction)} \tag{2.14}$$

The complication arises with the following electron transfer, which may proceed via two possible routes depending on whether the formation of the H–H bond occurs concurrently with the second electron transfer (Heyrovsky reaction) or whether dimerisation occurs between two *H species after the electron transfer has already taken place (Tafel reaction):

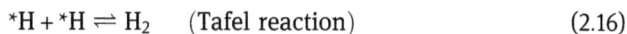

$$^*H + H^+ e^- \rightleftharpoons H_2 \quad \text{(Heyrovsky reaction)} \tag{2.15}$$

$$^*H + {}^*H \rightleftharpoons H_2 \quad \text{(Tafel reaction)} \tag{2.16}$$

Which of these reaction routes is followed will depend on the nature of the catalyst and reaction conditions employed. Computational studies by Skúlason et al. indicate that the mechanism on the Pt(111) facet follows the Volmer–Tafel route, since this is the lowest energy route [99]. However, we can see that the Tafel reaction requires two *H bound on neighbouring catalytic sites. This is not always possible on certain catalyst surfaces, such as for single-atom catalysts or in the presence of strongly bound poisoning species on the catalyst surface. Targeting surface sites on the catalyst to hinder the Tafel mechanism is therefore a common means to slowing the rate of HER in order to improve the faradaic efficiencies for all CO_2RR products.

3 Catalyst design

3.1 Carbon monoxide and syngas

Gold catalysts: Au is one of the most active materials for CO_2 reduction to CO but Au is limited by low abundance and high cost. It is often used as a model catalyst for comparing electrode or reactor design components. In all cases, the key motivation is to gain the maximum catalytic activity through the lowest Au mass. Mariano et al. used a combined bulk and scanning electrochemical microscopy to show that grain boundaries at Au surfaces enhanced CO_2RR to CO, but had no impact on the water of water reduction, offering to control the grain boundary density as a means to enhance the faradaic efficiency [100].

Substantial enhancement of Au catalysts can be achieved through nanostructuring or dispersion of nanoparticles on a high surface area, conductive support [101]. Targeted nanostructuring can be highly precise, specifically engineering nanoparticles to reveal active crystal facets. Lee et al. demonstrated this through the synthesis of complex concave rhombic dodecahedral Au nanoparticles, which produced a number of high index facets such as (331), (221) and (553) (Figure 3.1) [102]. Zhu et al. showed a similar effect on ultrathin Au nanowires, giving 94% CO at 8.16 mA cm^{-2}, and a high pass activity of 1.84 A g$_{Au}^{-1}$. The authors attributed this activity to a high density of reactive edge sites that also show very weak CO binding to favour product release [103].

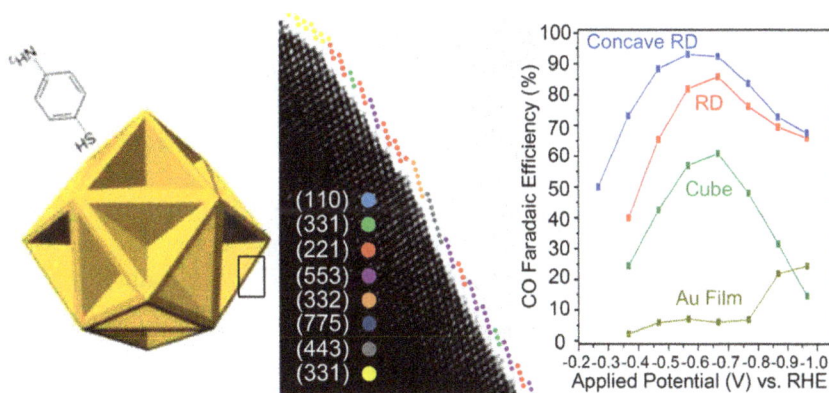

Figure 3.1: Schematic diagram for a concave rhombic dodecahedral (RD) Au nanoparticle for CO_2 reduction to CO. The templating method produces a series of highly active facets, shown in the central image. Activity for the RD nanoparticle exceeds other Au nanoparticle designs when operated under the same conditions. Adapted with permission from reference [102], Copyright © 2015 American Chemical Society.

Silver catalysts: Ag provides a cheaper alternative with a good selectivity towards CO but is limited by high overpotentials and a tendency for catalytic deactivation

https://doi.org/10.1515/9781501522239-003

leading to HER to be favoured over CO_2RR [49]. A number of groups have therefore worked to develop new micro- and nanostructured Ag materials in order to produce more low coordination, highly active CO_2RR sites at the catalyst surface [104]. Further improvements come from dispersing Ag onto an active, high surface area support such as multiwalled carbon nanotubes, which offered an impressive 95% faradaic efficiency of up to 350 mA cm^{-2} [105].

Ag and Au alloys and hybrids: A number of alternative catalysts have been designed based on Ag or Au metal alloys. These have the advantages of lowering the precious metal content as well as fine-tuning the surface properties of the catalyst. Nanostructured AgCu and AgInCu foam electrodes can be prepared either directly via electrodeposition [106] or through metal displacement on existing Cu foams [107]. Although their current densities are low due to their use in liquid-phase H-cells, the techniques for manufacturing foam electrodes facilitate relatively simple upscaling.

Gao et al. prepared a hybrid Au–CeO$_x$ catalyst by co-deposited Au and CeO$_x$ nanoparticles onto a carbon support. The hybrid catalyst outperformed the individual Au/C and CeO$_x$/C materials, giving 89.1% CO at 12.9 mA cm^{-2}. The authors used DFT to propose that the Au–CeO$_x$ interface strengthened the adsorption of *COOH to give a lower energy CO_2RR route [108]. Kim et al. demonstrated the same effect at AuCu nanoparticles, where bonding between O of *COOH and Cu in the alloy surface increases its stability [109]. This is an example of breaking the scaling relationship, since *COOH is stabilised without having a detrimental impact on *CO.

Cheaper alternatives: The cost and scarcity of Ag and Au led to researchers investigating a wide range of alternative CO_2RR catalysts for CO production. Zinc has been a popular choice; various nanostructuring techniques have employed to enhance its activity, usually starting from bulk Zn foil as the core material [110, 111]. One example of Zn nanowires reached a %FE of 98% at 40 mA cm^{-2} [112]. Similarly, some Ni catalysts have shown promising properties for CO_2RR to CO. One example on an activated carbon fibre support reached 67% efficiency for CO at up to 63 mA cm^{-2} [113]. Although both of these values are lower than those reported for Au and Ag, the cost and scarcity advantages of Zn and Ni mean that these numbers are still encouraging.

Single-atom catalysis: Advances in these cheaper materials focus on improving the catalytic activity while simultaneously hindering parasitic hydrogen evolution. Li et al. demonstrated this by developing a single-atom Ni catalyst with Ni confined in Ni–N$_4$ units in an N-doped carbon support. This provided high numbers of Ni active sites while hindering particle aggregation to give 99% CO at 28.6 mA cm^{-2} (Figure 3.2) [114]. Single-atom catalysts have added benefits of hindering hydrogen evolution, since single metal sites do not provide adjacent sites needed for *H dimerisation according to the Tafel mechanism. Instead, hydrogen evolution must proceed via the Volmer–Heyrovsky mechanism, which has much higher activation energy barrier, so the overall rate is decreased [115].

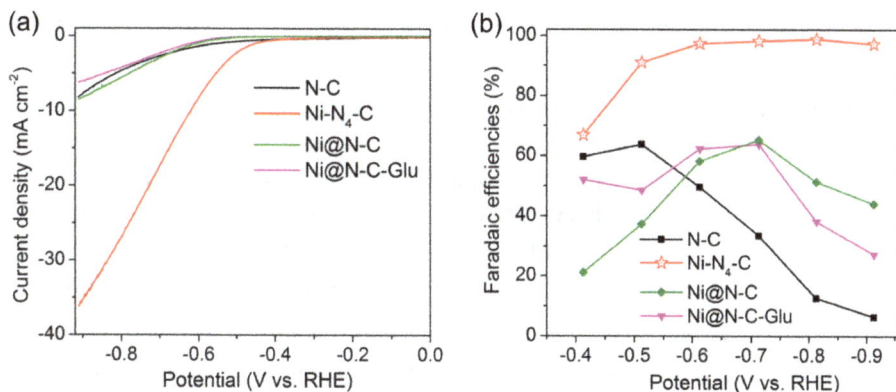

Figure 3.2: Linear sweep voltammograms (a) and faradaic efficiencies for CO (b) for N-doped carbon (N–C), single-atom Ni in N-doped carbon (Ni–N$_4$–C) and standard Ni/N–C catalysts made by pyrolysing Ni-doped g-C$_3$N$_4$ (Ni@N–C) and Ni-doped g-C$_3$N$_4$ with glucose (Ni@N–C–Glu). Ni–N$_4$–C outperforms all materials in terms of current density and faradaic efficiency. Adapted with permission from reference [114], Copyright © 2017 American Chemical Society.

Single-atom Ni catalysts have been demonstrated on graphene [116] and carbon nanotube supports [117], representing a promising route to high-activity, low-cost catalysts. Other single-atom catalysts have also been investigated. Bi et al. found that replacing Ni with other transition metals such as Fe or Co decreased the activity, which was suggested to be due to stronger CO binding poisoning the active metal sites [116]. Studies from Yang et al. and Zhang et al. produced Zn–N$_4$ [118] and Bi–N$_4$ [119] based materials, respectively, which gave up to 95% CO. However, the current density was lower than the Ni examples at ~ 5 mA cm^{-2} in both cases.

Palladium hydride: An interesting alternative material is Pd. Although Pd itself tends to favour formate production, applying a negative potential under aqueous conditions leads to the formation of palladium hydride (PdH), which is an active material for CO formation. The PdH surface has a much lower CO binding energy, favouring CO release as a dominant product [120]. Lee et al. showed that alloying Pd with a transition metal can further increase the amount of CO produced, with PdAg performing best out of the studied materials [121].

Metal chalcogenide catalysts: Other material options focus on metal chalcogenide materials. Although their low conductivity hinders their use as a bulk material, a number of works have successfully produced CO from metal chalcogenide nanoparticles on conductive carbon supports. Qin et al. produced 92% CO for CdS on carbon nanotubes, although the poor conductivity required a larger overpotential compared to leading metallic catalysts [122]. Other groups have doped metal chalcogenides with CO$_2$RR active materials to increase their performance, such as $Zn_xCd_{1-x}S$ [123] and Cu/In_2O_3 [124], where the level of doping has a sizeable impact on the CO yield versus H_2 as the other major product.

Syngas: Research into the targeted electrochemical production of syngas (CO and H_2 mixtures), as opposed to pure CO, has favoured different operating conditions. The similarity of the standard potentials for CO and H_2 production means that some degree of H_2 formation is inevitable during CO production. For this reason, syngas research tends to focus on producing materials and reaction conditions that reliably produce a targeted ratio of $CO:H_2$, based on the requirements of the syngas end use (Figure 1.3).

Tuning the $CO:H_2$ ratio: The simplest means of controlling the $H_2:CO$ ratio is through the applied overpotential. H_2 is produced from water, which is abundant at the electrode surface, whereas CO is produced from CO_2, which has a limited availability due to its solubility and diffusion rate. For this reason, as the applied overpotential is made more negative, the syngas ratio tends to favour H_2 over CO as the dominant product. Unfortunately, widespread production of tailored syngas through an optimised overpotential is not feasible. Larger overpotentials are needed in order to give fast production rates and industrially relevant product quantities. This presents a challenge to produce syngas ratios that favour CO while operating at high reaction rates.

Impact of particle size: To date, only a few studies exist that comprehensively investigate the impact of changing operating conditions or catalyst materials design on the ratio of $CO:H_2$ from a CO_2RR reactor. Mistry et al. demonstrated the impact of changing the size of Au nanoparticles on the production of syngas from CO_2RR (Figure 3.3). They showed that decreasing the particle size from ~ 8 to ~ 1 nm resulted in a significant increase in current density and simultaneous shift in $H_2:CO$ to favour H_2 [125]. They ascribed this to the formation of more low-coordinated Au sites on smaller nanoparticles that favour H_2 production. A similar effect was seen by Zhu et al. for CO_2RR at Au nanowires [103], and by Jeon et al. at Zn nanoparticles [126], where a decrease in the critical dimension increased the amount of H_2 in the product stream.

Changing catalyst active sites: Tuning active sites that favour either HER or CO_2RR to increase H_2 or CO component has also been achieved by alternative means. Marques Mota et al. varied the loading of Au nanoparticles on titanate nanoparticles, where increasing the Au content increased the CO production, with H_2 as the only other product detected [127]. Qin et al. achieved control over the $H_2:CO$ ratio by modulating the ratio of (002) and (101) crystal facets at Zn catalysts, where (101) favours CO and (002) favours H_2. Tuning the crystal facets allowed them to detect $H_2:CO$ ratios between 1:2.31 and 5:1 [128].

The $H_2:CO$ ratio has also varied via defect engineering at catalyst surfaces. Qin et al. showed that the proportion of CO produced versus H_2 is increased at CdS nanoparticles as the number of S-vacancies is increased [122], and Geng et al. showed the

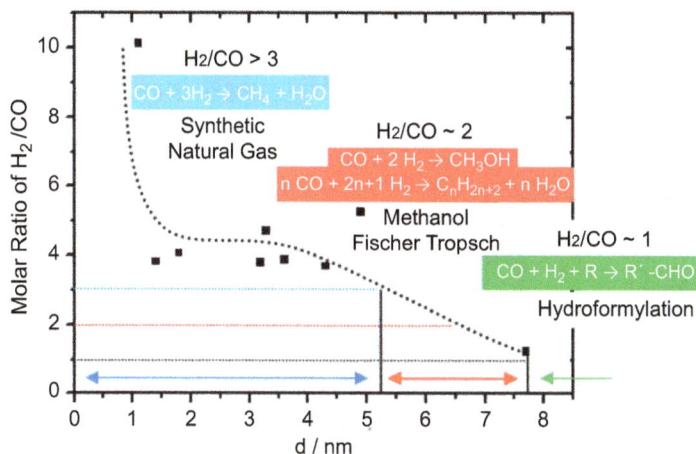

Figure 3.3: Molar ratio of the volume percentage of H_2 versus CO as a function of nanoparticle size recorded during the CO_2RR at Au nanoparticles in 0.1 M $KHCO_3$ at -1.2 V versus RHE. Potential applications of the resultant $H_2:CO$ ratio are shown in the coloured boxes. Adapted with permission from reference [125], Copyright © 2014, American Chemical Society.

same trend as the number of O-vacancies increased in ZnO nanosheets [129]. Daiyan et al. demonstrated a similar impact of increasing the defects in carbon catalysts, where the progressive removal of nitrogen from N-doped mesoporous carbon created an increasingly defect-rich surface that favours CO production [130].

3.2 Formate and formic acid

Tin catalysts: The starting point for electrochemical CO_2RR to formate catalysis is Sn, which sits at the peak of the volcano plot for formate selectivity (Figure 2.3). A simple Sn foil electrode in a liquid H-cell has been reported to produce ~70% formate at -7.3 mA cm^{-2} [70]. Multiple groups have worked to improve on this by introducing various micro- or nanostructuring techniques to maximise the activity of the Sn catalyst. Lei et al. reached 90% formate at 10 mA cm^{-2} with a Sn nanoparticle catalyst [131]. More complex structures have allowed these high efficiencies to be recorded at even higher current densities. Li et al. produced mesoporous SnO_2 nanosheets on carbon cloth that gave 87% formic acid at 50 mA cm^{-2} [132].

Tin alloy catalysts: Alloying Sn with additional elements can also alter the catalyst surface to favour formate production at higher current densities. A wide variety of different alloys have been investigated, including Ag_3Sn [133], $In_{0.9}Sn_{0.1}$ [134], BiSn [135] and CuSn [136]. Choi et al. demonstrated a series of SnPb alloys that outperformed both individual metals for formate production. The efficiency was dependent on the degree of doping, with $Sn_{56.3}Pb_{34.7}$ giving 79.8% formate at 57.3 mA cm^{-2} [137]. Bai et al. showed a similar dependence for the Pd:Sn ratio in PdSn alloys, where a 1:1 ratio gave almost 100% formic acid measured at a relatively low overpotential (Figure 3.4) [138].

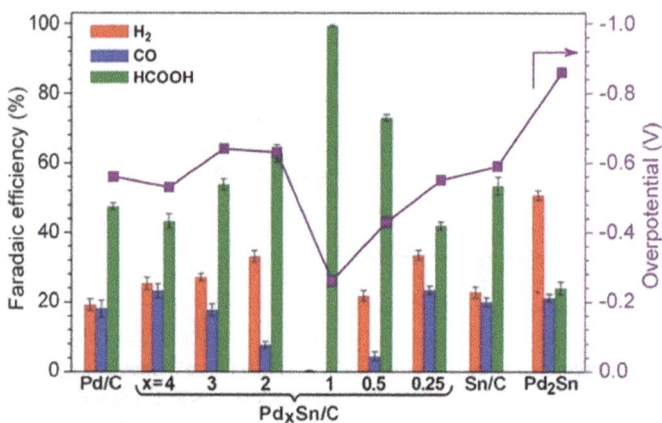

Figure 3.4: Faradaic efficiency (left y-axis) and overpotential (right y-axis) for Pd/C, Sn/C and Pd$_x$Sn/C catalysts for CO$_2$RR in CO$_2$-saturated 0.5 M KHCO$_3$. The alloy PdSn/C can be seen to give almost complete production of formic acid with negligible CO or H$_2$. Reproduced with permission from reference [138], Copyright © 2017 Wiley-VCH Verlag GmbH & Co. KGaA, Weinheim.

Defect engineering: Formate yield can also be enhanced by engineering specific defects in the catalyst surface, either to enhance the CO$_2$RR or to hinder HER. Ye et al. found that a SnCu alloy gave 82.3% formate at 79 mA cm^{-2}, thanks to a favourable formation of the stepped (211) surface, which is selective towards formate production over HER [139]. Luc et al. engineered a complex core–shell catalyst made up of a AgSn alloy core with a thin SnO$_x$ shell, which gave ~ 80% formate at ~ 20 mA cm^{-2} [133]. The authors used DFT to correlate the activity to oxygen vacancies in the SnO(101) surface.

Alternative pure metal catalysts: Although the literature is dominated by Sn and Sn alloys, other catalysts have been investigated for formate production via CO$_2$RR, mostly centred on other p-block materials such as Pb, In and Bi. In all cases, a number of different methods have need used to produce high surface area materials with a large number of highly active sites. Pander et al. used oxide- and sulphide-derived Pb to access high surface area Pb structures, with the sulphide-derived material producing 88% formate at ~ 13.6 mA cm^{-2} [140]. Luo et al. electrodeposited In onto a Cu mesh to access a hierarchically porous In material, which gave ~ 90% formate at 67.5 mA cm^{-2}. Zhang et al. showed that high surface area Bi nanosheets significantly outperformed bulk Bi for CO$_2$RR to formate, reaching 92% formate at ~ 3.7 mA g$_{Bi}^{-1}$ [141].

Outside of p-block elements, Co primarily produces a mixture of CO and formate [142], so some works have made modifications to alter the Co surface to shift the selectivity towards formate. Gao et al. demonstrated that oxygen vacancies at oxidised Co surfaces can stabilise formate intermediates [143], allowing them to produce partially oxidised ultrathin Co layers that gave ~ 90% formate at 10 mA cm^{-2} [144].

Single-atom catalysts: As discussed for CO in Section 3.1, formate production has been successful at single-atom catalysts, which provide an active and dispersed catalytic centre to give high CO_2RR activity while simultaneously hindering HER [115]. Zu et al. produced a single-atom Sn catalyst on N-doped graphene that gave 74.3% formate at ~ 12 mA cm^{-2} (Figure 3.5) [145]. The authors attributed this activity to charge delocalisation from Sn into the N-doped carbon support giving a positive charge on the catalytic Sn ($Sn^{\delta+}$), which favours CO_2 adsorption and also stabilises the CO_2^{-} radical formed after the first electron transfer. Taheri et al. also produced up to 96% formate at a single-atom Fe catalyst, although this was done at a lower current density of 4 mA cm^{-2} [146].

Doped catalyst materials: Alternatively, materials can be enhanced through doping in order to modulate the reactions at the catalyst surface. Ma et al. improved the formate production of an In catalyst through sulphur doping, where the S site produced surface-bound hydrogen species that were then readily available to react with CO_2 to give formate [147]. The S-doped In gave 85% formate at 95 mA cm^{-2}, which is particularly impressive for a liquid-phase cell. Sulphur has also been used to enhance formate production at Cu [148] and Sn [149] catalysts, giving 74% formate at 14.5 mA cm^{-2} and 93% formate at 55 mA cm^{-2}, respectively.

Increasing the current density: The rate of CO_2RR to formate in liquid H-cells is inherently limited by the low solubility of CO_2 in aqueous electrolytes. In order to overcome this limitation and reach higher operating current densities, it is necessary to move to gas diffusion electrodes (GDEs). These move CO_2 mass transport into the gas phase and help to overcome CO_2 solubility issues. A full discussion of GDEs for CO_2RR is given in Section 5.1. In fact, even a simple GDE with a Sn powder catalyst has been reported to produce ~ 90% formic acid at up to 200 mA cm^{-2}, once factors such as catalyst loading were optimised [150].

Tin versus tin oxide: Further advancements in GDE catalysts proceed in much the same way as for liquid-phase catalysts, focussing on surface modifications and structural changes to enhance the activity of the catalyst surface. Sen et al. compared the performance of Sn versus SnO_2 at GDEs (Figure 3.6). They showed that SnO_2 was selective over a wider potential range than Sn, and that the SnO_2 was kinetically more favourable for formate production [151].

ATR-IR spectroscopy studies suggest that the improved performance of SnO_2 is due to the presence of Sn^{2+} moieties at the oxide surface, which favourably bind to CO_2 and carbonate-like intermediate species involved in the CO_2RR

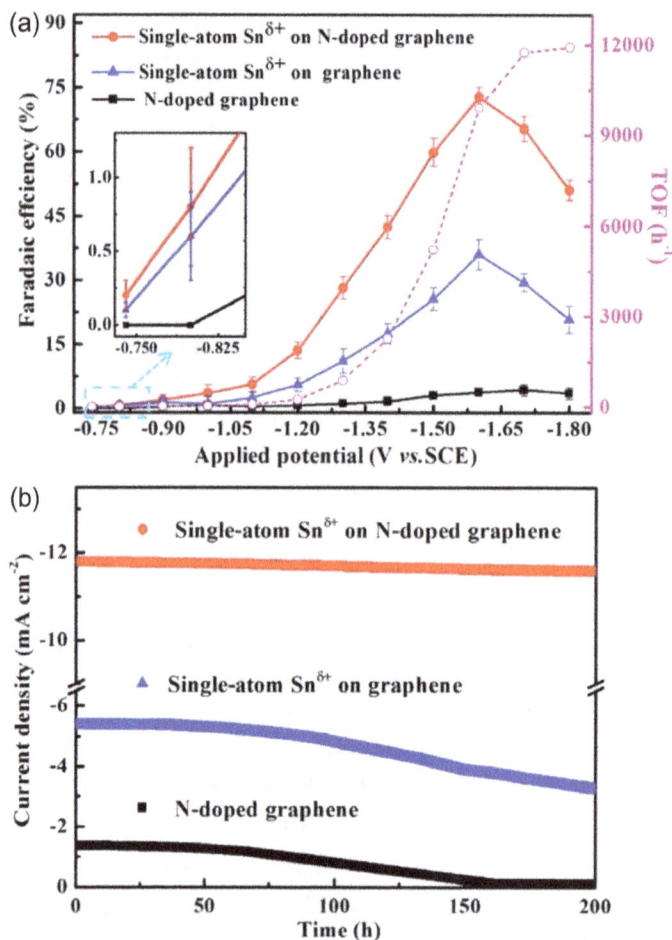

Figure 3.5: (a) Faradaic efficiencies for formate production in 0.25 M KHCO$_3$ at N-doped graphene (black), single-atom Sn$^{\delta+}$ on graphene (blue) and single atom Sn$^{\delta+}$ on N-doped graphene (red). Secondary y-axis shows the turnover frequency (TOF) for Sn$^{\delta+}$ on N-doped graphene at the corresponding potentials. (b) Current density over a 200 h constant reduction experiment. The N-doping can be seen to sizeably increase both the faradaic efficiency and stability, indicating that the N-doping in necessary to both activate and stability the Sn$^{\delta+}$ sites. Figure adapted with permission from reference [145], Copyright © 2019 Wiley-VCH Verlag GmbH & Co. KGaA, Weinheim.

mechanism [152, 153]. Recently, Qin et al. used this as a starting point to develop one of the best performing Sn-based CO$_2$RR catalysts for formate with SnO nanorod@nanosheet assemblies (nanosheets grown on the outside of nanorods), which produced an impressive 94% formate at ~ 330 mA cm^{-2} [154]. The authors attributed high performance to a combination of the availability of Sn^{2+} along with the prevalence of highly active, low-coordinated sites at the edges of nanosheets [155].

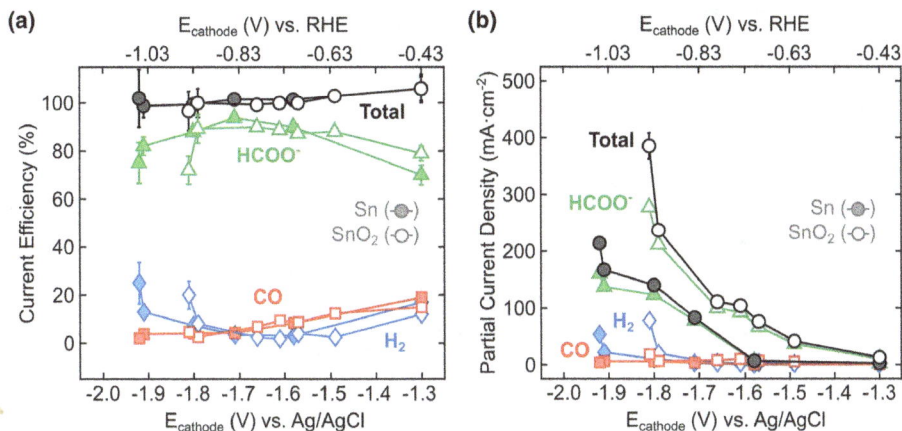

Figure 3.6: Plots of current efficiency (a) and partial current density (b) for Sn (filled symbols) and SnO$_2$ (hollow symbols) GDEs for CO$_2$RR in 0.5 M Na$_2$CO$_3$ and 0.5 M Na$_2$SO$_4$. Each data point corresponds to a 1 h reduction experiment. Partial current densities are the product of the total current density and the current efficiency recorded for each separate potential. Although the current efficiencies are similar, SnO$_2$ can be seen to give a much higher partial current density for formate. Reproduced from reference [151], Copyright © 2019, Springer Nature B.V.

3.3 Copper catalysts

A cursory search of the CO$_2$RR literature reveals a field dominated by copper and copper alloy materials. Copper is a pretty unique CO$_2$RR catalyst for two main reasons: (i) it is the only material capable of producing significant amounts of C$_{2+}$ CO$_2$RR products and (ii) it is capable of simultaneously producing a massive number of different CO$_2$RR products. It is worth noting that copper is not the only metal capable of this behaviour, as is discussed in Section 2.2, but copper is certainly the only material that reliably produces C$_{2+}$ products with impressive faradaic efficiencies. This latter point was recently highlighted by Jaramillo et al., who reported the simultaneous detection of 16 different CO$_2$RR products from CO$_2$RR at a copper electrode, although some were detected with faradaic efficiencies as low as 0.1% (Figure 3.7).

This unique nature of Cu in terms of its CO$_2$RR selectivity presents a certain challenge in describing its use as a CO$_2$RR catalyst. Cu has been used as a starting point for catalytic research into methane, methanol, ethylene and ethanol. In virtually all cases, where one of these four species is the primary target, the other three are also formed to some extent. The key for developments in CO$_2$RR at Cu catalysts is therefore to shift the selectivity towards one specific product, either by favouring that particular route or by hindering all of the others.

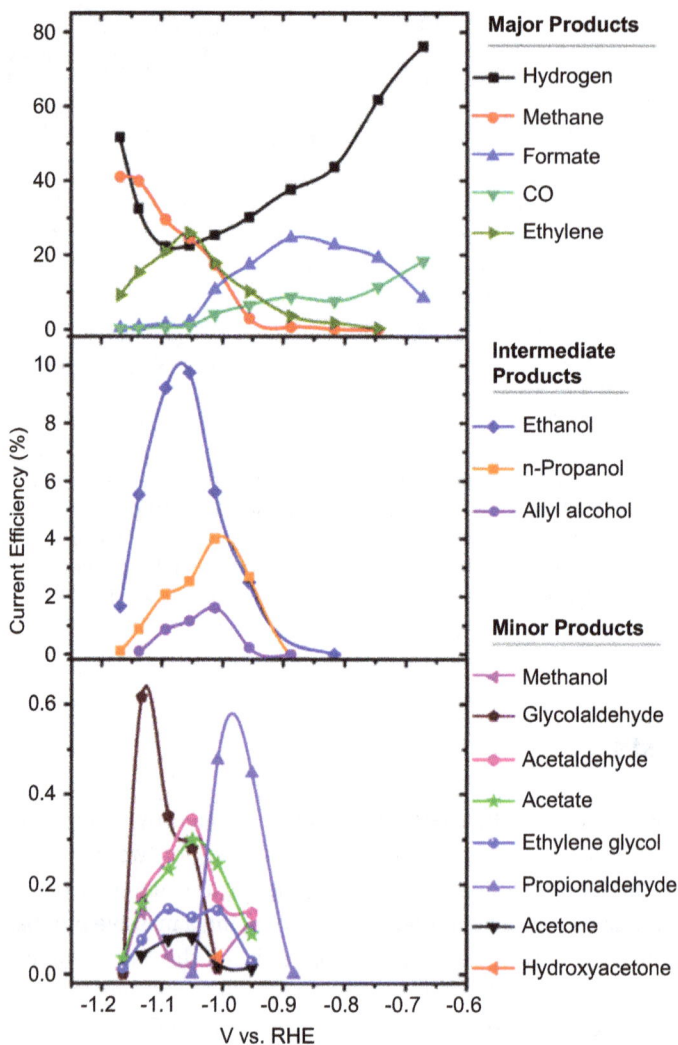

Figure 3.7: Faradaic efficiency for 16 different CO_2RR products that were detected during CO_2RR at a copper foil electrode in a flow cell configuration. Reproduced with permission from reference [90], Copyright © 2012 RSC Publishing.

C_1 versus C_2 selectivity: Broadly speaking, differentiation between C_1 products (methane and methanol) and C_2 products (ethylene and ethanol) centres on favouring or disfavouring the formation of the C–C bond. Chapter 2 highlighted that the choice of crystal facet has a significant impact on the stability of different intermediates in the CO_2RR mechanism. These differences in stability are evidenced by a clear change to produce selectivity across different Cu facets.

Hori et al. demonstrated this concept at Cu single-crystal electrodes. For CO_2RR at 5 mA cm^{-2}, Cu(100) preferentially formed ethylene, whereas Cu(111) preferentially formed methane [156]. As well as favouring the appropriate intermediates, the square geometry of the Cu(100) facet favours *CO dimerisation to give the C–C bond [157]. The focus on favouring or disfavouring C–C bond formation is a recurring theme when looking at Cu-based materials for C_1 and C_2 products.

3.3.1 Methane and methanol

Cu nanostructures: Selective nanostructuring techniques can be employed to reveal specific active sites that favour methane or methanol over ethylene or ethanol. Wang et al. used crystal etching on Cu nanocubes to reveal the high-energy (110) facet, which increased the methane selectivity from ~ 45% to ~ 58% versus the starting Cu(100) nanocubes [158]. Interestingly, even the non-etched nanocubes showed more methane than ethylene despite the dominance of the (100) facet, which may indicate an impact of the Pd core upon which the Cu nanocube was originally grown.

A similar effect has been observed at polycrystalline Cu nanoparticles. Manthiram et al. used a highly dispersed Cu nanoparticle catalyst on carbon support, with particles ~ 25 nm in diameter to produce 76% methane at ~ 13 mA cm^{-2} [159]. The authors proposed that the high dispersion is necessary to reveal active crystal facets, whereas high nanoparticles hide these facets as particles aggregate. Li et al. revealed a highly methane-selective Cu surface by synthesising fivefold Cu nanowires (Figure 3.8). This gave a high number of active under-coordinated sites, providing 55% methane at ~ 12 mA cm^{-2} [160]. They further enhanced the catalyst stability by wrapping the wires in graphene, which hindered particle dissolution and sintering that has been found to alter the size and shape of Cu nanostructures [159].

Cu oxide: Cu oxide surfaces have been employed to shift the CO_2RR selectivity towards methanol. Le et al. found that Cu oxide thin-film electrodes outperformed anodised or air-oxidised equivalent Cu electrodes, suggesting that the stability of Cu(I) at the electrode is key for methanol production [161]. Azenha et al. produced Cu nanowires from CuO rather than metallic Cu. They found that the oxidised surface hindered HER and gave a favourable surface for methanol as the primary product, giving 66% methanol at 45 mA cm^{-2} [162].

Single-atom catalysts: A number of groups have taken the approach of isolating Cu catalytic sites so that CO_2 intermediates are held far apart in order to drive selective C_1 formation by hindering C–C bond formation. One of the simplest ways to provide the activity of Cu without facilitating C–C bond formation is to use a highly dispersed Cu catalyst on a support that is nor particularly active for CO_2RR. Zhao et al. produced a highly dispersed oxide-derived Cu/C catalyst by pyrolysing a Cu-MOF

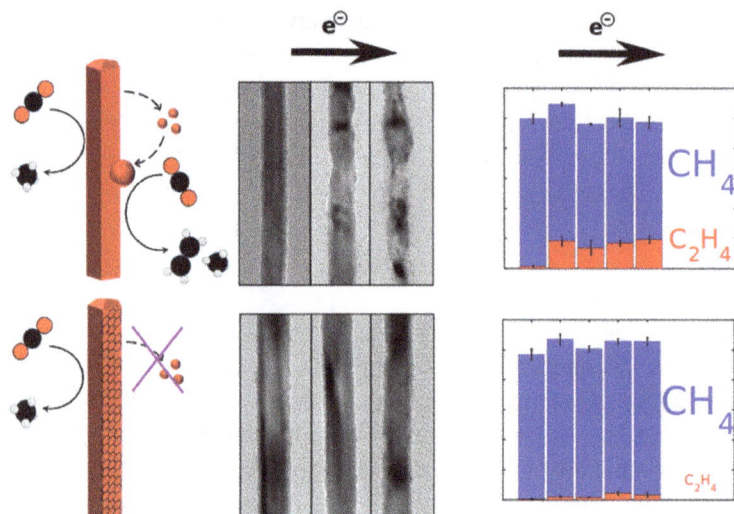

Figure 3.8: (Left) Schematic diagrams for fivefold Cu nanowires for CO_2RR. The pentagonal cross section can be seen at the top of both wires. The bottom wire is graphene wrapped to prevent Cu dissolution or re-deposition. The centre and right columns show SEM images of the wires and the evolving faradaic efficiencies of methane and ethylene, respectively, over a period of extended CO_2RR operations. The graphene layer maintains the nanowire structure, whereas the bare wire undergoes morphology changes that lead to more ethylene being produced over time. Reproduced with permission from reference [160], Copyright © 2017 American Chemical Society.

(metal organic framework). The resultant material showed good selectivity towards alcohols, giving ~ 43% methanol and ~ 27% ethanol [163].

Even greater dispersion can be achieved with single-atom Cu catalysts, which can be produced by a number of different routes. Wang et al. used single-atom Cu on CeO_2 nanorods (Figure 3.9). Cu content was kept low at 4% to provide isolated Cu coordinated to oxygen vacancies, which could then produce 58% methane at 30 mA cm^{-2} [164]. Yang et al. produced a single-atom Cu catalyst dispersed on carbon nanofibers that gave up to 44% methanol at 93 mA cm^{-2}, with the rest being CO [165]. Tan et al. confined single Cu atoms in an MOF to shift the selectivity of Cu from ethylene to methane [166]. Construction of Cu-MOF around a Cu_2O aided charge transfer and produced a high surface area, resulting in 63% methane detected at ~ 13 mA cm^{-2}.

Cu alloys: An alternative approach is to alloy Cu with a secondary metal. Tuning the level of the alloying element can impact the likelihood of C–C bond formation and also impact the stability of surface-bound intermediates. Takatsuji et al. produced a series of CoCu alloys, where increasing Co content gave formate at low overpotentials and methane at more negative overpotentials as Co hindered *CH_2 dimerisation and also facilitated hydrogenation of *CH_2 to *CH_3 [167]. Zhou et al. used a CuFe alloy combined with a silicon photoelectrode to give up to 51% methane at

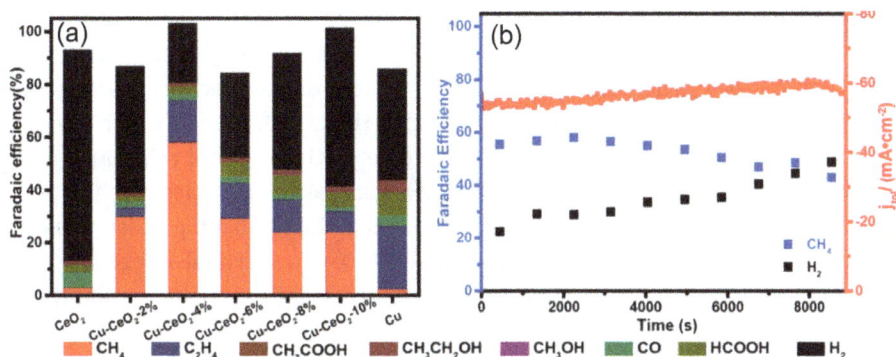

Figure 3.9: (Left) Faradaic efficiencies for a number of CO_2RR products recorded at Cu–CeO$_2$ nanorod catalysts with varying levels of Cu doping. The percentage of Cu has a sizeable impact on the product distribution, with 4% Cu giving the most methanol. (Right) Stability for the 4% Cu–CeO$_2$ catalyst over prolonged CO_2RR conditions. The faradaic efficiency can be seen to be relatively consistent over 2 h of operation. Adapted with permission from reference [164], Copyright © 2018 American Chemical Society.

38 mA cm^{-2} [168]. The authors used DFT to demonstrate that the binary CuFe distorts CO_2 on adsorption to lower the energy barrier for reduction to methane.

Wang et al. analysed a number of CuBi nanoparticle alloys that showed a strong compositional dependence on the methane selectivity, with Cu$_9$Bi giving up to 70.6% methane at ∼ 54 mA cm^{-2} [169]. The authors proposed that the high selectivity comes from the electronegative Bi can give a partially oxidised Cu site, increasing its activity. Jia et al. found a similar compositional dependence for CuAu alloy nanoporous films, with Cu$_{63.9}$Au$_{36.1}$ giving 15.9% methanol along with 12.1% ethanol and 12.6% formate [170].

3.3.2 Ethylene and ethanol

Copper: Copper is unique in its ability to produce sizeable amounts of C$_{2+}$ CO_2RR products, with ethylene being the dominant C$_{2+}$ product in virtually all cases. The ethylene selectivity of simple Cu foil can reach up to 20% ethylene at relatively low current densities [171]. Experimentally, it has also been observed that C–C bond formation is favoured at more under-coordinated sites [172]. Hori et al. demonstrated this by introducing 110 and 111 steps into a Cu(100) surface. By tuning the ratio of step to plane sites, they were able to increase the ethylene:methane ratio from 0.2 at Cu(111) up to ∼ 10 at Cu(711) [173].

Copper nanoparticles and nanostructures: Employing Cu nanoparticles and nanostructures provides access to larger numbers of under-coordinated sites to increase the ethylene yield, as well as increasing the electrochemically active surface area.

Tang et al. demonstrated this by varying the treatment of polycrystalline Cu surfaces, with Cu nanoparticles outperforming electropolished and argon-sputtered catalysts [174]. The Cu structure can also be tailored to produce specific crystal facets. For example, Cu nanocubes preferentially give the Cu(100) facet, which favours dimerisation to give C_2 products [175]. Loiudice et al. found that Cu nanocubes also show size-dependent ethylene selectivity, with 44 nm particles reaching 41% ethylene at mA cm^{-2} (Figure 3.10) [176]. Roberts et al. found that the onset potential for Cu nanocubes was less negative than the Cu(100) single crystal, suggesting that the nanocube structure provides a greater enhancement than can be simply explained by its dominant crystal facet.

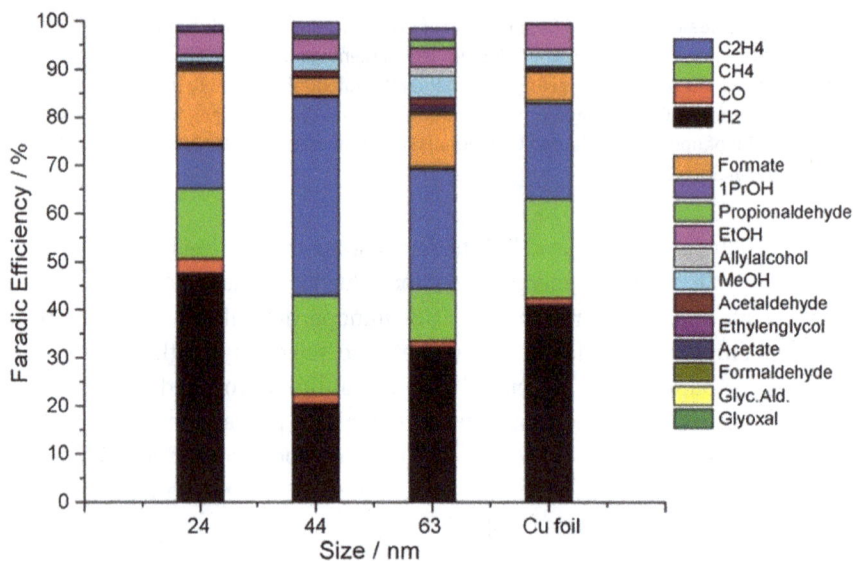

Figure 3.10: Faradaic efficiencies for a number of possible CO_2RR products formed at Cu nanocubes of varying sizes at −1.1 V versus RHE in 0.1 M $KHCO_3$. About 44 nm particles are optimal for ethylene formation as well as for reducing parasitic H_2 production. Reproduced from reference [176], Copyright © 2016 Wiley-VCH Verlag GmbH & Co. KGaA, Weinheim.

Wang et al. used electrodeposition under CO_2RR conditions to produce catalytically active Cu surfaces. Interestingly, they found that performing the electrodeposition while simultaneously reducing CO_2 caused a capping effect that preferentially created Cu(100) facets [177]. The catalyst provided an impressive 60% ethylene and up to 90% C_{2+} selectivity at 350 mA cm^{-2}.

Other nanostructuring techniques have focused more on creating highly active under-coordinated sites rather than high proportions of the Cu(100) facet. Reller et al. produced highly active copper nanodendrites by dissolving and redepositing Cu in acidic solutions [178]. The resultant electrode gave 57% ethylene at 170 mA cm^{-2}, although the structures suffered from stability losses due to sintering and oxidation processes. Hoang et al. used a 3,5-diamino-1,2,4-triazole inhibitor in a Cu electroplating bath to give a high surface area film that gave 40% ethylene and 20% ethanol at ~ 150 mA cm^{-2} [179].

The surface structure also has an impact on the output ratio of ethylene:ethanol. Computational studies by Zhuang et al. suggested that vacancy sites on a Cu surface increase the energy barrier to ethylene but do not impact the route to ethanol [180]. Based on this, they synthesised nanoparticles with a Cu_2S core and vacancy-rich Cu shell, which gave 24.7% ethanol and 21.2% ethylene, compared to 7.5% ethanol and 45.2% ethylene at the equivalent Cu catalyst.

Oxide-derived copper: One of the most popular means to simply access Cu nanostructures is through the reduction of Cu oxides. The oxidation/reduction cycling can be done electrochemically, with the precise cycling conditions used offering a high level of control over the nature of the resultant structure. Reller et al. produced highly active copper nanodendrites by dissolving and redepositing Cu in acidic solutions [178]. The resultant electrode gave 57% ethylene at 170 mA cm^{-2}, although the structures suffered from stability losses due to sintering and oxidation processes. Kwon et al. used cyclic voltammetry in halide solutions to a similar effect. Sweeping the potential positively gave cubic Cu_2O nanocrystals, which were then reduced on the negative sweep to give ~ 20 nm Cu nanoparticles with irregular dimensions, giving 50% more ethylene versus polycrystalline Cu [181].

Despite numerous works demonstrating highly efficient C_{2+} CO_2RR product generation at oxide-derived catalysts, there is still much debate regarding the origin of this increased efficiency [182]. Correlative studies have highlighted multiple changes in the nature of oxide-derived copper versus equivalent metallic materials. Ren et al. found that CO_2RR on Cu_2O thin films gave up to 39% ethylene and 16% ethanol [183]. They ascribed the high selectivity to the formation of sub-micron metallic Cu polyhedron particles at the electrode surface during reduction, which provided the ideal crystal facets for C_2 formation. Tang et al. proposed that the increased activity of their Cu nanostructures was due to the presence of more active under-coordinated sites [174].

Mistry et al. compared the activity of oxygen and hydrogen plasma-treated Cu (Figure 3.11). They showed that samples with large numbers of Cu(I) sites gave more ethylene than equally rough surfaces without Cu(I) [184]. The faradaic efficiency was proportional to the degree of oxygen plasma treatment, reaching a maximum ~ 60% ethylene at ~ 12 mA cm^{-2}. Xiao et al. suggested that oxide-derived Cu produces a mixed matrix of Cu(I) and Cu(0), which facilitates the dimerisation of adsorbed *CO to give C_2 products [185]. Liang et al. designed a Cu_3N catalyst so that the nitride base would stabilise Cu(I) sites, which was able to produce 39% ethylene and 15% ethanol at ~ 22 mA cm^{-2} [186].

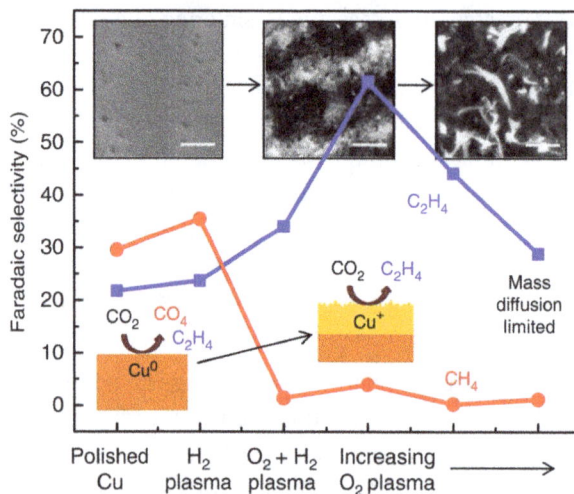

Figure 3.11: Faradaic efficiency for methane (red) and ethylene (blue) at Cu surfaces from varying degrees of plasma treatment. Insets show SEM images for H_2 plasma-treated (left), low O_2 plasma-treated (middle) and high O_2 plasma-treated (right) Cu foils. Scale bars represent 500 nm. O_2 plasma treatment gives a sizeable increase in ethylene selectivity but further oxidation systematically lowers this efficiency. Reproduced with permission from reference [184], available open access on Creative Commons license CC BY, Copyright © 2016 Mistry et al.

Lum et al. suggested that the increased surface area achieved with oxide-derived catalysts was able to generate a locally high pH at the electrode surface [82]. Since high concentrations of OH^- hinder H_2 evolution and high pH C–C bond formation at lower overpotentials, the net effect is an increased C_{2+} yield. They do note, however, that targeting an increased local pH is a compromise between favouring C–C bond formation while still giving sufficient availability of CO_2 at the electrode surface. As discussed in Section 1.3.2, CO_2 is tied up in an aqueous equilibrium with bicarbonate and carbonate anions. High concentrations of OH^- can react with CO_2 to form bicarbonate or carbonate, which lowers CO_2 availability for electrochemical reactions. The authors observed this for an oxide-derived catalyst with a roughness factor (R_f = electrochemically active area/geometric area) of 103, where the CO_2 availability became so low that the dominant cathodic product was hydrogen.

Copper halide catalysts: Similar to Cu_2O as a route to catalytically active Cu(I) sites, multiple groups have reported improved C_2 product yields at Cu surfaces modified with halides. Yano et al. created Cu-halide mesh electrodes by oxidising the Cu mesh in the halide potassium salt. CO_2 was then reduced in that same electrolyte, with the CuBr mesh electrode giving 79.5% ethylene and 1.6% ethanol at ∼ 40 mA cm^{-2} [187]. Ma et al. investigated Cu surfaces modified with F, Cl, Br and I [188]. All of the studied materials provided larger C_{2+} faradaic efficiencies with respect to bare Cu, with the F-modified Cu performing best with ∼ 51% ethylene and ∼ 17% ethanol at 1.6 A cm^{-2}.

Copper alloys: The ability of Cu surfaces to favour C–C bond formation means that CO_2RR to C_2 products is not as commonly studied at alloy surfaces as is done for other metals. Section 3.3.1 highlighted how tuning the alloying components of a Cu alloy can inhibit this ability to shift selectivity towards C_1 products. Ma et al. demonstrated this at a series of CuPd alloy catalysts (Figure 3.12) [189]. Even dispersion of Cu and Pd throughout the catalyst hindered dimerisation, so CO was the dominant product, whereas phase-separated Cu and Pd gave ~ 48% ethylene and ~ 15% ethanol at 360 mA cm^{-2}.

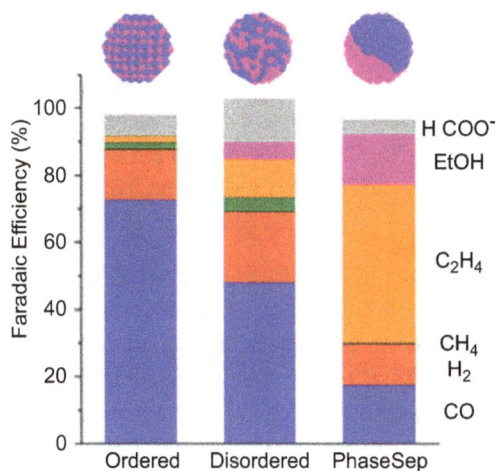

Figure 3.12: Faradaic efficiencies for multiple CO_2RR at CuPd nanoparticle catalysts. Three configurations of mixing within the alloy particles are shown; ordered (left), disordered (centre) and phase separated (right). Greater degrees of mixing reduce the number of neighbouring Cu sites, which hinders C–C bond formation to give a lower proportion of C_2 products. Reproduced with permission from reference [189], Copyright © 2017, American Chemical Society.

Sinton et al. found that the proportion of Pd in a CuPd alloy also impacted the ethylene:ethanol ratio, since Pd sites provided active *H at their surface which favoured the hydrogenation steps in the ethanol mechanism [78]. CO_2RR at a CuPd$_{0.007}$ gave 33.2% ethanol and 37.0% ethylene at ~ 700 mA cm^{-2}, compared to 20.1% ethanol and 54.8% ethylene at an equivalent Cu catalyst. Interestingly, when Chen et al. doped Cu$_2$O with PdCl$_2$, they found that the selectivity shifted away from ethylene to give 30.1% ethane at 19.5 mA cm^{-2} [190]. The authors suggested that the PdCl$_x$ sites were highly active in the hydrogenation of C_2H_4 intermediates so that the fully hydrogenated ethane was formed instead.

Ren et al. took a similar approach with an oxide-derived Cu$_x$Zn catalyst, where the ethylene:ethanol ratio was dependent on the Zn content [191]. The Cu$_4$Zn catalyst gave 29.1% ethanol and 10.8% ethylene at ~ 30 mA cm^{-2}, compared to 11.3% ethanol and 26.5% ethylene at the equivalent Cu material. A combined computational and experimental approach by Li et al. showed that alloying Cu with the more weakly binding Ag could shift the selectivity towards ethylene, since intermediates on the ethylene pathway were destabilised while those on the ethanol pathway were unaffected [192]. Optimising the Ag:Cu ratio produced 41% ethanol at 250 mA cm^{-2}, compared to 29% ethanol at an equivalent Cu electrode.

3.3.3 Acetate and acetic acid

The mechanism for acetate formation via CO$_2$RR is very similar to that of ethanol, with the key requirements being the formation of the C–C bond and maintaining C–O bonds within the intermediates as additional electron transfers occur. As such, much of the references to acetate within the CO$_2$RR literature are as a by-product in the targeted electrosynthesis of ethanol or ethylene. The CuPd alloy developed by Li et al. for ethanol selectivity also produced up to 14.9% acetate at ~ 700 mA cm^{-2} [78].

Surface structure: The surface dependence of the acetate versus ethanol selectivity means that even slight modifications to the nature of the Cu surface can give a sizeable shift in the acetate:ethanol ratio. Zhu et al. created a series of oxide-derived Cu catalysts, where Cu was dissolved, coordinated to an organic ligand, then redeposited and electrochemically reduced (Figure 3.13) [193]. By changing only the choice of ligand, they were able to cause enough of a structural change to vary the acetate selectivity between 17.5% and 48.5% at 11.5 mA cm^{-2} on equivalent electrodes.

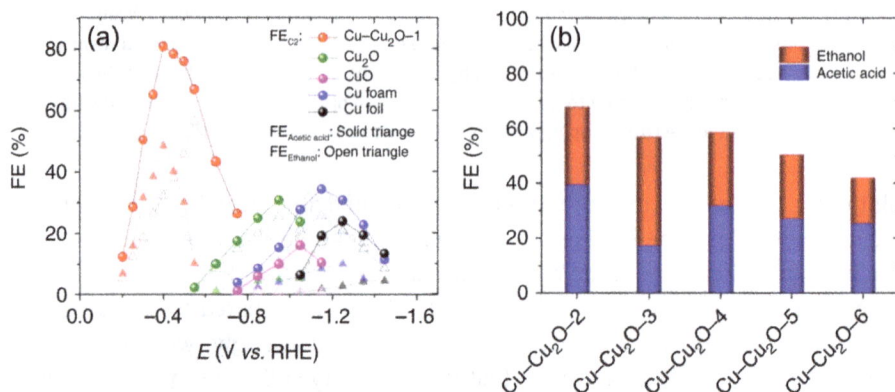

Figure 3.13: (a) Faradaic efficiencies (FE) for C$_2$ products (solid circles), acetic acid (solid triangle) and ethanol (open triangle) at different Cu surfaces. The oxide-derived Cu–Cu$_2$O–1 catalyst gives improved faradaic efficiencies for all species at less negative potentials. (b) Faradaic efficiencies for ethanol (red) and acetic acid (blue) at different oxide-derived catalysts. The precise surface structure gives a sizeable shift in the relative selectivity towards the two products. Adapted with permission from reference [193], available open access on Creative Commons license CC BY Copyright © 2019, Zhu et al.

Role of the carbon support: Similar to variances in nanostructure, substrate modifications can shift the selectivity towards or away from acetate. Genovese et al. found that dispersing Cu nanoparticles on carbon nanotubes gave 56.3% acetic acid, along with 35.4% formic acid and 8.3% methanol at 100 mA cm^{-2} [194]. The authors ascribed the high activity to the high rate of CO_2^- formation, which reacted with $*CH_3$ at the Cu surface to drive C–C bond formation onto acetate as the dominant product.

Wang et al. produced a Cu/N-doped nanodiamond catalyst, where Cu nanoparticles were sputtered over an N-doped nanodiamond support, which gave 34% acetate and 28.9% ethanol at ~0.7 mA cm^{-2} [195]. Interestingly, a small amount of acetate was also detected at the bare N-doped nanodiamond but not at pure carbon nanodiamond. The authors proposed that the N-doped nanodiamond played a role in driving C–C bond formation with Cu sites, hence the high selectivity for C_2 oxygenates.

3.3.4 Propanol

Access to C_3 products follows a similar trend to C_2 in that the literature is dominated by Cu and Cu alloy catalysts, with propanol being the most commonly observed. The key focus in catalytic materials is providing the advantageous surface features that drive the formation of C–C bonds while holding the reaction intermediates at the electrode long enough for a second C–C bond to form, but not making the binding so strong that the surface can become poisoned by CO or other adsorbates. As a result, the highest concentrations of C_3 are most often detected at the more complex Cu structures. As mentioned in the previous section for acetate, much of the literature for propanol electrosynthesis reports it as a by-product of ethylene or ethanol production. This means that faradaic efficiencies for propanol are low even at very low current densities when combined with the inherent difficulty in driving multiple C–C bond formations to the same intermediate.

Nanostructured surfaces: Zhuang et al. achieved this by creating Cu surfaces with nanocavities, which trapped C_2 intermediates at the electrode surface to drive coupling reactions between C_1 and C_2 to give C_3 (Figure 3.14) [196]. By tuning the geometry of the nanocavities through acid-etching Cu_2O, they were able to produce 21% propanol at 7.8 mA cm^{-2}. Kim et al. produced an assembly of differently shaped Cu nanoparticles, where the initial electrolysis caused Cu morphology changes to give a mixture of nanocubes and smaller nanoparticles [197]. The resultant surface gave 5.9% propanol, 27% ethylene and 13.3% ethanol at 12.8 mA cm^{-2}.

Oxide-derived Cu: As observed for C_2 products, oxide-derived Cu has been shown to produce some of the largest proportions of C_3 products, although the exact mechanism of this enhancement is still debated. Lee et al. produced a chloride-induced oxide-derived Cu catalyst [198]. The Cl$^-$ stabilised Cu_2O units in the catalyst to give a biphasic Cu–Cu_2O structure, producing 7.8% propanol, as well as 23% ethylene,

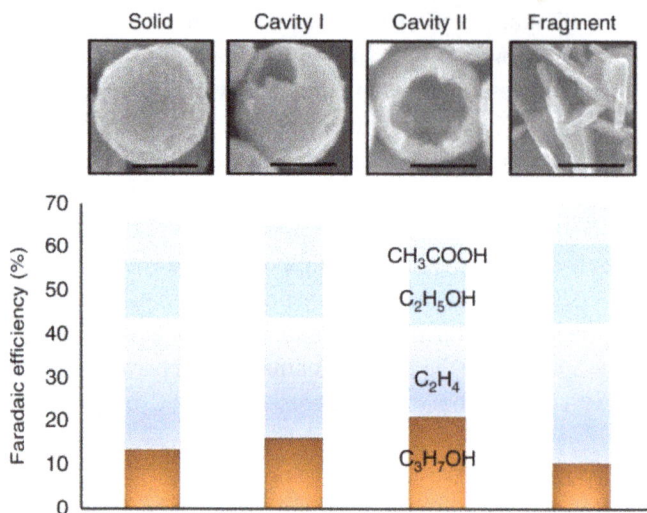

Figure 3.14: Faradaic efficiency for the detected C_2 and C_3 CO_2RR product at different Cu nanoparticle morphologies as measured at −0.56 V versus RHE. The corresponding nanoparticle structure is shown in the SEM image above each stacked bar. Scale bar is 100 nm. Adapted with permission from reference [196], Copyright © 2018 Zhuang et al. under exclusive licence to Springer Nature Limited.

20% ethanol, 1% propane and 1% butane. Ren et al. electrochemically reduced a Cu_2O film to give Cu nanocrystal agglomerates. The oxidation/reduction treatment revealed various crystal orientations and grain boundaries, producing 8.75% propanol, 35.8% ethylene and 12.7% ethanol at 19.9 mA cm^{-2} [199]. Kim et al. created oxide-derived Cu nanowires in order to give an active Cu mesostructured [200]. The availability of Cu features of multiple different length scales produced 12.4% propanol with 10.6% ethylene and 9.2% ethanol at 6.8 mA cm^{-2}.

Cu alloys: Chen et al. doped an oxide-derived Cu catalyst with $PdCl_2$ in order to create $PdCl_x$ islands. These Pd sites produced large quantities of *H at the catalyst surface to drive the hydrogenation of bound intermediates, giving 5.5% propanol, along with 30.1% ethane and 11.1% ethanol at 19.5 mA cm^{-2} [190]. Li et al. also detected propanol at their $CuPd_{0.007}$ alloy catalyst, although the faradaic efficiency was ~2% lower than the equivalent Cu catalyst at 6.3% as the modifications were aimed at promoting ethanol yield [78]. Ren et al. varied the Zn content in an oxide-derived CuZn film. The best performing Cu_4Zn gave 5.39% propanol with 10.7% ethylene and 29.1% ethanol at 37.3 mA cm^{-2} [191].

3.4 Tandem catalysts

Tandem catalysts work on the principle that a catalyst can be produced with two different types of site that have distinct roles in the overall mechanism. In the case of CO_2RR, tandem catalysts provide an opportunity to access higher order reduction products, namely methanol, methanol, ethylene and ethanol, by combining Cu catalytic sites with a second material that can enhance its selectivity or activity as needed. Primarily, there are two types of tandem catalysts that can achieve this based on their mode of operation.

CO formation at tandem sites: The most commonly employed tandem catalyst focuses on the initial reduction of CO_2 to CO. The methodologies detailed earlier in this chapter for producing C_1 products from electrochemical CO_2 reduction should not be thought of as the end of the story. C_1 products formed through the CO_2RR are not necessarily inert. This is especially important when considering one of the most common CO_2RR products: CO. CO itself is electrochemically active. A number of works have demonstrated that CO feed-gas for electrochemical reactors can be used for the production of value-added carbon products such as methane and ethylene in much the same way as CO_2 [201]. Romero Cuellar et al. demonstrated the concept by hyphenating two catalysts in series, where CO_2 was first reduced to CO at Ag electrode, then further reduced to C_2 and C_3 products at Cu electrode downstream [202].

Based on this, it is possible to envisage alloying Cu with a material that can efficiently perform the initial reduction of CO_2 to CO via a two-electron reduction reaction. CO would then be transferred to the Cu site for further reduction to the desired end product. This often discussed as a means of breaking the scaling relationship between *CO and *CHO, as discussed in Section 2.3. The initial reduction of CO_2 to CO would be done at a material with an optimised *CO, then the second material that drives the further reduction of *CO to the end product would have a separately optimised *CHO [96]. Using two different materials means that the catalysis is not limited by the inherent linearity between *CO and *CHO as if the same material were used for both stages. O'Mara et al. described this as a "nanozyme", analogous to an enzyme capable of multistep cascade reactions [204]. Based on this they produced a catalyst with an Ag core and porous Cu shell, so that CO_2 is reduced to CO at the Ag core, then further reduced to a mixture of C_2 and C_3 products in the Cu shell.

Tandem alloy catalysts: Huang et al. used this concept for an AgCu alloy catalyst, where the presence of the Ag component to facilitate the initial two-electron reduction gave a 3.4-fold increase in the ethylene faradaic efficiency versus an equivalent Cu catalyst (Figure 3.15) [203]. The role of Ag in providing CO to Cu sites was demonstrated by Hoang et al. [205]. When feeding a Cu catalyst with CO_2 they observed a plateau in the faradaic efficiency for ethylene due to a limited supply of CO for further reduction. In the presence of Ag, this plateau was not present. Equally, when CO was supplied in the gas flow, the plateau also disappeared, indicating that Ag is continually supplying CO

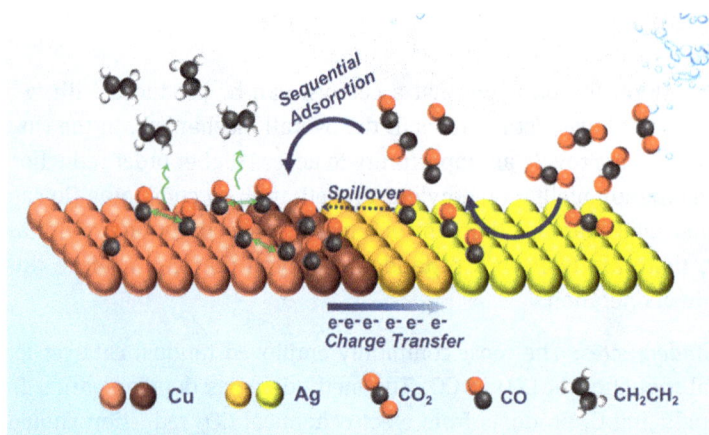

Figure 3.15: Schematic diagram for a tandem AgCu catalyst for ethylene-selective CO_2RR. CO_2 is first reduced at the Ag site to give CO, which is then passed to the Cu site for further reduction to ethylene. Reproduced with permission from reference [203], Copyright © 2019 American Chemical Society.

to the Cu sites to facilitate CO_2RR (Figure 3.16). The net result was a catalyst providing 55.2% ethylene and 25.9% ethanol at 310 mA cm^{-2}. Xiong et al. developed a highly efficient Ag/Au nanoframe catalyst for CO_2 reduction to CO. When the nanoparticles were decorated with Cu, the catalyst became highly selective for ethylene, giving up to 77% ethylene at 235 mA cm^{-2} [96].

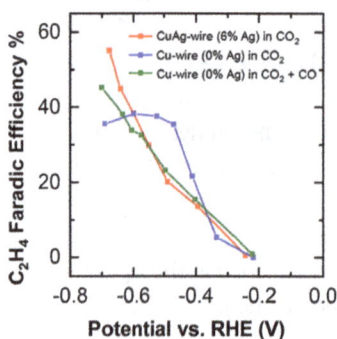

Figure 3.16: Faradaic efficiency for ethylene formation from a Cu catalyst with CO_2 (blue), Cu catalyst with CO_2 and CO (green) and CuAg alloy with CO_2 (red). The similarity in trend between the Cu catalyst in the presence of CO_2 and CO and the CuAg catalyst in the presence of CO_2 only indicates that the Ag is providing CO to Cu sites for further reduction. Reproduced with permission from reference [205], Copyright © 2018 American Chemical Society.

Feng et al. demonstrated a similar effect at CuZn catalysts, where Zn supplied CO to the Cu site for further reduction, although the resultant selectivity was lower, giving 33% ethylene at 6.1 mA cm^{-2} [206]. Moving to CuO/ZnO nanoparticles on a carbon support allowed Li et al. to use the same concept at the oxidised catalyst surface, giving 50.9% ethylene and 22% ethanol at 367 mA cm^{-2} [207].

Tandem catalysts for C$_1$ species: Tandem catalysts can also be employed for methane using this concept. The challenge here is for *CO transfer to Cu to accelerate *CO reduction without driving the formation of C–C bonds. Zhang et al. achieved this by decorating Ag electrode with dispersed Cu sites [208]. They carried out CO$_2$RR at an Ag foil with 1.5 ppm Cu in the electrolyte, giving a sparsely decorated surface with Cu. Optimising the Cu loading allowed them to produce ~ 60% methane at ~ 8.3 mA cm^{-2}.

Providing reactive hydrogen: Alternatively, the secondary site is designed to facilitate a subsequent reaction after CO is formed. This favours the further reduction of CO to a bound intermediate before it can be released as the primary product. In these cases, the alloying element is usually chosen to provide reactive *H in close proximity to a Cu site with a bound *CO. Zhang et al. found that optimising the Pd: Cu ratio to Cu$_2$Pd favoured surface Pd–H reacting with *CO, giving up to 48% methane at ~ 2.5 mA cm^{-2} [209]. Guo et al. saw a similar effect at a Cu$_3$Pt catalyst, giving around ~ 21% methane at ~ 5 mA cm^{-2} [210].

3.5 Non-copper catalysts for higher order products

Although the field is clearly dominated by Cu and Cu alloy materials, methane, methanol, ethylene and ethanol have been detected at other catalyst surfaces during CO$_2$RR (Figure 3.17). In most cases, these are detected at very low faradaic efficiencies and are mentioned only as an unwanted by-product rather than as the focus of the study. Methane and methanol are detected more often than ethylene and ethanol, most likely because they do not require C–C bond formation, for which Cu is so well suited. In fact, to date methane and methanol have been detected, albeit in low amounts, at Zn, Ag, Ni and Pt as well as Cu [211].

C$_1$ products: Some C$_1$ products have been detected at surprisingly high faradaic efficiencies on non-Cu materials. In these cases, significant modifications have been made to the catalyst structure in order to maximise its activity. Wu et al. produced a Co porphyrin immobilised on carbon nanotubes capable of producing ~ 40% methanol at ~ 25 mA cm^{-2} [212]. The authors state that the high efficiency for methanol is only possible because the Co porphyrin is bound to nanotubes as individual molecules, whereas Co porphyrin aggregates on the same support would produce methanol with a much lower activity and selectivity. Huang et al. produced nanoparticles

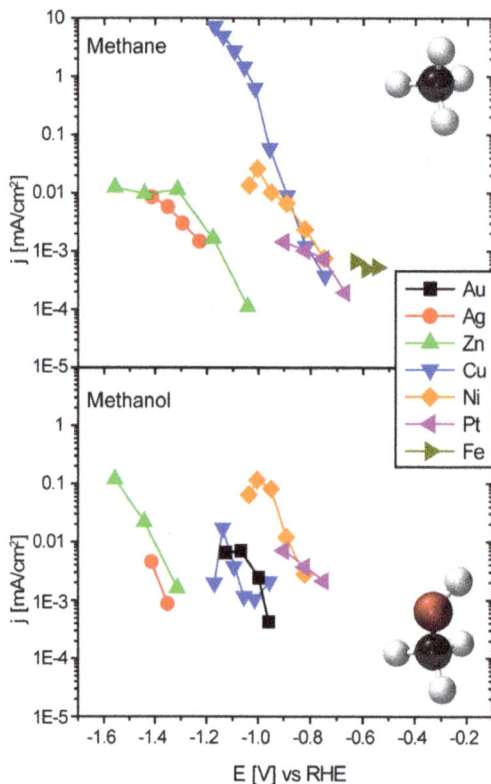

Figure 3.17: Partial current densities for methane and methanol during CO_2RR at a number of different metals. Reproduced with permission from reference [211], Copyright © 2014 American Chemical Society.

of $Co(CO_3)_{0.5}(OH) \cdot 0.11H_2O$ with high surface area "urchin-like" structures [213]. These gave an impressive 97% methanol, although the current density was low at 0.59 mA cm^{-2}.

Zhang et al. developed a complex catalyst from Pd/SnO_2 nanosheets, which gave 54% methanol at 1.45 mA cm^{-2} [214]. Capping Pd with SnO_2 favoured CO_2 adsorption while also weakening the binding strength of CO in order to prevent surface poisoning of CO sites. Umeda et al. were able to produce methane at a Pt/C catalyst by lowering the CO_2 partial pressure in the gas stream [215]. Although the faradaic efficiency was low at 6.8%, the required overpotential was very low, which the authors ascribed to the prevalence of surface-bound *H at the surface that is involved in the CO_2RR mechanism. Qu et al. produced up to 60.5% methanol using RuO_2/TiO_2 nanotubes at ∼ 1 mA cm^{-2} [216]. The nanotubes significantly outperformed the equivalent nanoparticles, indicating a strong structural dependence.

Han et al. were able to produce methane at a single-atom Zn catalyst on a porous N-doped carbon support, giving 85% methane at ∼ 37 mA cm^{-2} [217]. DFT calculations suggested a preferential bond between Zn and O in CO_2, which favoured CH_4 generation of CO as the dominant product. The ability of single-atom Zn to produce methane was also exploited by Lin et al., who used a tandem catalyst where Co porphyrin oxidised CO_2 to CO, which was passed to Zn on N-doped carbon for reduction to methane. The Co porphyrin was also able to increase the availability of *H for the methane mechanism, giving 18.3% methane at ~ 240 mA cm^{-2}.

C_{2+} *products:* Non-copper catalysts for C_{2+} products are fewer and further between. Kortlever et al. proposed that C_{2+} formation at non-Cu surfaces proceed via the polymerisation of *CH_2 at the catalyst surface [91]. Based on this they designed a PdAu alloy catalyst to optimise the *CO binding energy, which was able to produce a large number of different C_{2+} species, including ethylene, ethane, propene, propane, butane, butane, pentane and pentene. Although the faradaic efficiencies were extremely low, their detection highlights that the line should not be drawn under Cu as the only intrinsically viable material.

Gonglach et al. produced a Co–corrole complex on carbon paper that gave 48% ethanol, along with 8% methanol and 10% acetate at ~ 2.5 mA cm^{-2} [218]. The authors proposed that the Co(III) site was able to stabilise the radical intermediates in the CO_2RR pathway. A similar Mn(III) corrole was more selective towards acetate, giving 63% acetate and 16% methanol at 0.8 mA cm^{-2} [219]. The C_2 selectivity was proposed to come from the Lewis acidic Mn binding of CO_2RR intermediates via O atoms, which facilitates CO dimerisation to oxalate, which is subsequently reduced to acetate.

Liu et al. were able to produce ethanol at a Ru polypyridyl carbene catalyst on N-doped carbon, giving 27.5% ethanol and 10.9% acetate at ~ 3 mA cm^{-2} [220]. CO_2RR at the Ru carbene itself results in CO as the only product, so the C–C bond formation must be due to a synergistic action between the Ru carbene and the N-doped carbon,

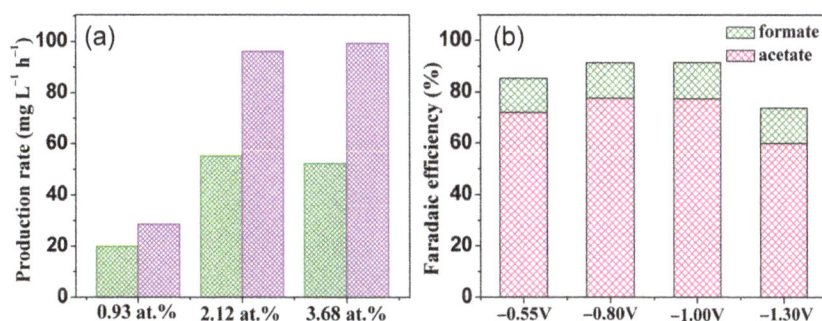

Figure 3.18: (a) Production rates for acetate and formate on N-doped nanodiamond arrays with varying levels of N-doping. The increasing N content gives a sizeable increase in the production of both species. (b) Faradaic efficiency for formate and acetate at the N-doped nanodiamond, showing a high selectivity over a broad range of potentials. Adapted with permission from reference [221], Copyright © 2015 American Chemical Society.

possibly due to the highly porous structure confining CO intermediates to facilitate CO–CO coupling reactions. Liu et al. achieved a similar effect at N-doped nanodiamonds on a Si rod array, which gave up to ~ 77% acetate and ~ 14% formate at ~ 7 mA cm^{-2} (Figure 3.18) [221]. Tafel analysis revealed a faster rate of acetate production versus formate production, suggesting C–C dimerisation was fast once the initial electron transfers occurred.

4 Electrolytes

Simple electrochemical systems can function to a high degree with relatively little concern for the choice of electrolyte, so long as the solution is sufficiently conductive to minimise the resistance and the electrolyte is relatively inert over the operating potential range. However, many real electrochemical systems can show drastically different behaviours depending on the choice of electrolyte employed. This is particularly prevalent for surface-dependent electrochemical processes; adsorption of the electrolyte onto the catalyst surface can block active sites for electrochemical reactions and different electrolytes can have stabilising or destabilising effects on the intermediates of multi-step electrochemical mechanisms. In the case of CO_2RR, the choice of electrolyte can not only impact the overall rate of reaction but the impact of electrolyte on the stability of surface-bound reaction intermediates means that an electrolyte can also impact the selectivity of a reaction towards a desired product.

4.1 Aqueous electrolytes

Aqueous electrolytes are arguably the most popular choice for CO_2RR reactors on small and large scales. All reaction mechanisms require proton transfers to take place, so aqueous media provides an abundant supply, which is particularly important for accessing higher order C_{2+} products. The desired conductivity is then achieved by dissolving a certain quantity of inorganic salt in solution. Aqueous solutions and most commonly employed salts are cheap and abundant and fairly easy to up-scale. However, as previously discussed, the CO_2RR reaction occurs outside of the water solvent window. All CO_2RR product streams invariably contain some degree of H_2 from the reduction of water at negative potentials. The choice of electrolyte is therefore defined not only by the impact of the electrolyte on CO_2RR mechanism, but often if it can slow the rate of parasitic water reduction.

4.1.1 pH

Slowing the HER: One of the most notable impacts of changes in pH is the impact it has on HER. Parasitic water reduction is a constant consideration for CO_2RR applications in order to minimise the quantity of H_2 in the gas outflow and improve the charge efficiency towards CO_2RR products. One of the simplest ways to lower the rate of H_2 evolution is to move to higher pH solutions. Alkaline electrolyte slows the kinetics of the first electron transfer step (Volmer reaction, eq. (2.14)), leading to a lower surface *H coverage and therefore a slower rate of H_2 evolution via the following electron transfer: Heyrovsky or Tafel reaction, eqs. (2.15) and (2.16) [222–224]. Dinh et al.

https://doi.org/10.1515/9781501522239-004

achieved some of the highest production rates for ethylene by operating a Cu GDE in highly basic 7 M KOH, giving a stable 70% ethylene at ~ 100 mA cm^{-2} for 150 h [225].

*CO dimerisation activity: High pH solutions have a secondary effect on the distribution of CO_2 in solution close to the electrode surface. In very basic solutions, CO_2 interaction with hydroxide leads to a lower penetration depth of CO_2 into the electrolyte, that is, CO_2 accumulates close to the electrode surface [226]. This higher availability of CO_2, combined with a reduced rate of HER as previously mentioned, leads to a greater accumulation of *CO at the electrode surface.

Further, DFT studies also indicated that the presence of hydroxide at the electrode surface lowers the *CO binding energy and stabilises the adsorbed *OCCO intermediate that forms as a result of *CO dimerisation. The net impact is seen as an increased rate of *CO dimerisation, increasing the proportion of C_{2+} species in the CO_2RR products. Dinh et al. used this concept when performing CO_2RR at Cu GDEs in varying concentrations of KOH. As the KOH concentration increased and the pH became more negative, CO_2 permeated less far into the solution, the rate of *CO dimerisation increased and the ethylene yield increased up to 66% at 275 mA cm^{-2} in 10 M KOH [225].

Product selectivity: The solution pH has a sizeable impact on the selectivity of CO_2RR at the Cu electrode where the product distribution is fairly broad. The predominant impact of pH is seen in the ratio of methane to ethylene seen in the outflow. Hori et al. showed that ethylene is favoured in KCl, K_2SO_4, $KClO_4$ and dilute $KHCO_3$, whereas concentrated $KHCO_3$ favoured methane [227]. They attributed this difference to the availability of protons at the electrode surface, which is controlled by the pH of the electrolyte. Varela et al. demonstrated the same effect using electrolytes with varying buffer capacities [228]. The rate of methane formation increased as the pH moved closer to neutral, although this came with a concurrent increase in H_2 formation by parasitic water reduction that is also favoured at lower pH (Figure 4.1).

In order to rationalise the impact of pH on the methane to ethylene ratio, it is necessary to consider the rate-determining step for the formation of each of these [229]. For ethylene, the rate-determining step has been proposed to be the formation of the C–C bond, either by the dimerisation of two neighbouring *CO molecules [81] or by the addition of *CO to an adsorbed *CHO [230]:

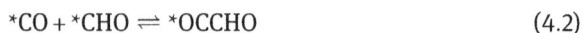

$$\text{*CO} + \text{*CO} \rightleftharpoons \text{*OCCO} \tag{4.1}$$

$$\text{*CO} + \text{*CHO} \rightleftharpoons \text{*OCCHO} \tag{4.2}$$

Neither of these steps involves a proton transfer, so the rate-determining step for C_2 product formation can be assumed to be independent of pH. On the other hand, the rate-determining step for methane formation has been proposed to be the hydrogenation of *CO to give *CHO [73]:

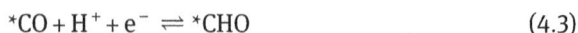

$$\text{*CO} + \text{H}^+ + \text{e}^- \rightleftharpoons \text{*CHO} \tag{4.3}$$

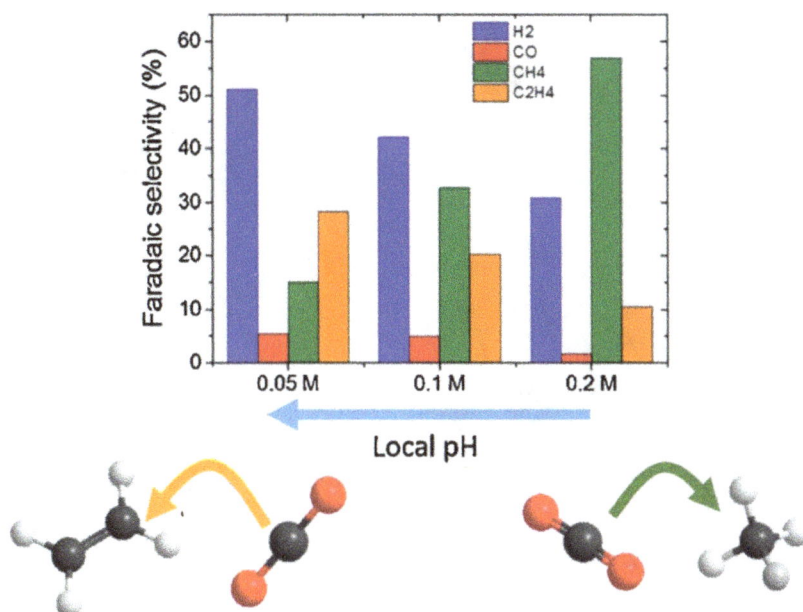

Figure 4.1: Impact of local pH on the selectivity of CO_2RR at a Cu electrode. As the buffer concentration decreases, the local pH becomes more alkaline, shifting selectivity towards ethylene. More concentrated buffers maintain a lower local pH, and so more methane is produced. Reproduced with permission from reference [228], Copyright © 2015 Elsevier B.V.

This reaction does involve a proton transfer, and therefore does display a pH dependence. An increase in pH therefore decreases the rate of *CO hydrogenation to slow the rate of methane production, while C–C coupling to give ethylene can still occur via the dimerisation of neighbouring *CO (Figure 4.2).

It is worth noting that ethylene production through the reaction between *CO and *CHO (eq. (4.2)) will display a pH dependence, since the initial formation of *CHO is pH dependent. This was observed by Schouten et al. for on Cu(111) surfaces, since the (111) facet favours ethylene formation via the *CHO intermediate [231]. The *CO dimerisation route is then favoured at the (100) facet. The influence of pH on methane versus ethylene yield at Cu electrodes would therefore have to take the proportion of the (111) facet versus (100) into account in order to determine which of the distinct pathways will define the surface reactivity.

Extremely high pH: The case for CO_2RR at extremely high pH presents an added layer of complexity. At very high pH, the overpotential required for methane formation and HER are both reduced, indicating a more kinetically facile route with commonalities between the methane formation and HER mechanisms [231]. One proposed reason for this is a switch in the formation of *H in the first HER step,

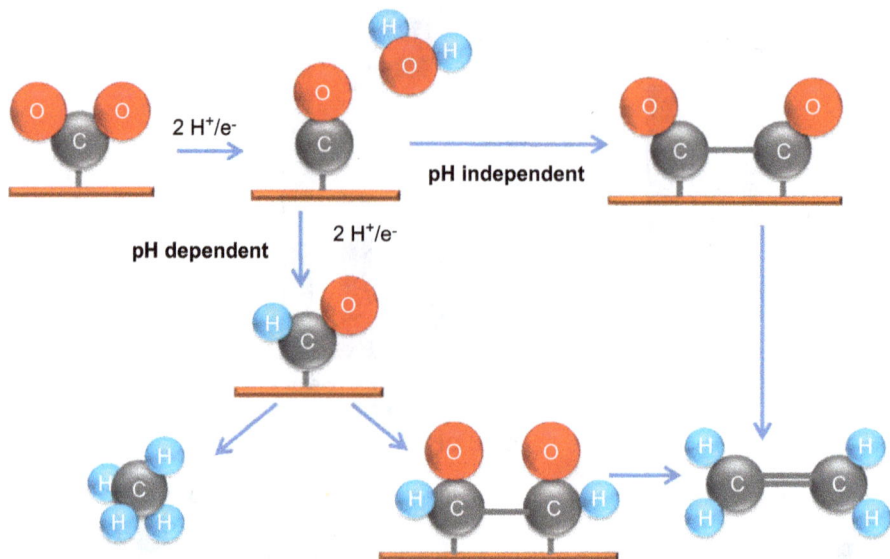

Figure 4.2: CO_2RR mechanisms for methane and ethylene. The dimerisation of *CO to give ethylene is pH independent, whereas the formation of either methane or ethylene via *CHO is pH dependent. Reproduced with permission from reference [228], Copyright © 2015 Elsevier B.V.

where the hydrogen source at high pH is from abundant water, rather than low availability of H^+ [232]:

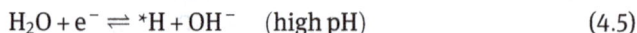

$$H^+ + e^- \rightleftharpoons {}^*H \qquad \text{(low pH)} \qquad (4.4)$$

$$H_2O + e^- \rightleftharpoons {}^*H + OH^- \quad \text{(high pH)} \qquad (4.5)$$

The change in mechanisms has been proposed to occur around pH 11–12. Importantly, the high pH route is independent of pH, and so may explain why both HER and *CO hydrogenation in the methane CO_2RR mechanism are accelerated at very high pH.

4.1.2 Anions

Bicarbonate: The most commonly reported electrolyte for CO_2RR studies is potassium bicarbonate ($KHCO_3$). HCO_3^- has a number of features that make it desirable as an electrolyte. As part of an equilibrium between HCO_3^- and CO_3^{2-}, bicarbonate has a pH buffering capacity [233]. This helps to minimise changes in pH during CO_2RR that are driving by the proton consumption involved in all CO_2RR mechanisms. Kinetic studies at Au electrodes have also indicated that HCO_3^- can act as a proton donor as part of the CO_2RR mechanism [234]:

$$CO_2 \rightleftharpoons HCO_3^- + H^+ \rightleftharpoons CO_3^- + 2H^+ \tag{4.6}$$

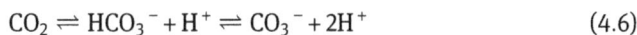

Perhaps most importantly is the impact that the bicarbonate equilibrium has on the CO_2 concentration in aqueous media. CO_2 solubility in aqueous solutions is unfortunately low. CO_2 depletion at the electrode surface is a key factor in many CO_2RR systems only producing high selectivity products at low reaction rates; attempting to increase the reaction rate when CO_2 is depleted will only accelerate the rate of water reduction and lower the charge efficiency due to H_2 production.

Isotopic labelling studies from Dunwell et al. have shown that HCO_3^- can act as a CO_2 donor in electrochemical experiments [235]. HCO_3^- decomposes to supply CO_2 to the electrode, and then further CO_2 from the gas flow can dissolve and replenish HCO_3^-. In fact, some groups have been able to produce CO_2RR products from the reduction of HCO_3^- itself, even in the absence of a CO_2 gas flow stream, by taking advantage of the $HCO_3^- \rightleftharpoons CO_2$ equilibrium [236, 237].

Halides: The CO_2RR environment is substantially different in halide solutions compared to HCO_3^-. It has already been discussed in Section 3.3 that halides can be added into the reaction environment around Cu electrodes in order to drive morphology changes. The solubilisation and re-deposition of Cu can create highly activity nanostructures, and the presence of halide ions has been reported to stabilise Cu(I) sites with high CO_2RR activity. Halides also have a number of impacts outside for driving morphological changes. It is worth noting that electrode surfaces will still be highly dynamic in halide solutions, and so studies probing one impact of a halide-based electrolyte may still be subject to the morphological changes caused by catalyst dissolution and re-deposition.

Halides are not involved in protonation equilibria and so have no buffering capacity. Hori et al. found that, for CO_2RR at copper electrodes, the selectivity was shifted towards ethylene in Cl^- solutions and towards methane in $KHCO_3$. The authors ascribed this to the increased local pH in unbuffered Cl^- [227].

A sizeable shift in product selectivity comes from the adsorption of halide anions on electrode surfaces. In the first instance, the adsorption of halide anions blocks active sites at the electrode surface. This can be advantageous since it blocks the surface sites needed to water reduction to hydrogen, and so the parasitic HER is limited. The degree of this effect is related to the strength of the binding interaction between the metal and the halide, where strongly binding I^- hinder reactions to a greater extent than more weakly binding F^- [238, 239]. The use of halides in this way facilitates CO_2RR in low pH solutions that would otherwise give impractically larger HER currents [240].

As well as impacting the HER, specifically adsorbed anions can facilitate the adsorption of CO_2 to the electrode surface (Figure 4.3). Ogura et al. suggested that electron transfer may occur from the adsorbed halide to the vacant orbital on CO_2 [239]. This results in a structural shift from linear to bent CO_2 and weakens that C–O bond, activating the CO_2 for the initial electron transfer. The more strongly adsorbing anions

Electrode

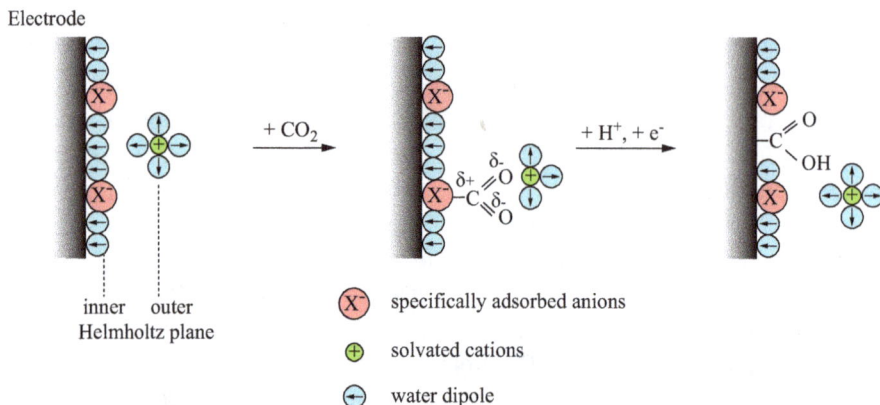

inner outer
Helmholtz plane

$\boxed{X^-}$ specifically adsorbed anions

$\oplus$ solvated cations

$\ominus$ water dipole

Figure 4.3: Schematic diagram for the mediation of CO_2 adsorption by the electrolyte within the electric double layer. CO_2 binds to the specifically adsorbed anion and the negative charge on the adsorbed CO_2 is stabilised by the cation. Reproduced with permission from reference [242], Copyright © 2013 Elsevier Ltd.

will donate more electron density onto the CO_2, giving a greater degree of activation. The resultant electron transfer can be thought of as a heterogeneous bridge electron transfer, where electrons move across the bridging halide ion between catalyst and CO_2 [241].

Varela et al. proposed a third impact of halides, where they could facilitate the hydrogenation of *CO at the catalyst surface [243]. This rationalised the authors' observations that I^- gave a greater increase in methane production versus ethylene production, since I^- could enhance the proton-dependent *CO hydrogenation but not the *CO dimerisation step.

4.1.3 Cations

Multiple studies have shown a significant impact of the cation choice on CO_2RR. A common theme is that the number of electron transfers can be increased by moving to larger cations. However, there is still some debate as the precise mechanism by which cations impact the CO_2RR reaction rate and/or its selectivity. The uncertainty seems to be centred on the degree of interaction between the cation and the electrode surface.

Specifically adsorbed cations: Murata et al. showed that the selectivity towards ethylene in bicarbonate electrolytes increased following the trend $Li^+ < Na^+ < K^+ < Cs^+$, with methane following the opposite trend [244]. The authors proposed that the more weakly hydrated Cs^+ ions adsorb more readily onto the electrode surface. This created a more positive effective potential difference between the electrode surface and the electrolyte, favouring the reduction of neutral CO_2 rather than the positively charged

H^+. The same effect was also seen by Paik et al., where formate production at an Hg electrode followed the trend $Li^+ < Na^+ < Et_4N^+$ [245].

Thorsen et al. also proposed that adsorbed Cs^+ from their electrolyte is responsible for the enhanced production of CO at an Ag electrode [246]. The authors suggested that the positively charged Cs^+ could stabilise the $*CO_2^-$ radical through ion pairing, thereby accelerating the overall CO_2RR to CO as the end product.

Cations in the Helmholtz layer: A number of other groups have stated that the cations do not specifically adsorb onto the electrode but stay hydrated in the outer Helmholtz layer close to the electrode surface [247, 248]. Instead, the influence of the positively charged cations comes from their position in the outer Helmholtz layer. Resasco et al. used DFT calculations to probe the interactions between adsorbed CO_2RR intermediates and various solvated cations (Figure 4.4) [249]. They showed that adsorbed $*CO_2$, $*CO$, $*OCCO$ and $*OCCOH$ species were stabilised via electrostatic interactions with the solvated cations. The trend of improved activity with larger cation size was attributed to larger cations being present in larger concentrations in the outer Helmholtz layer.

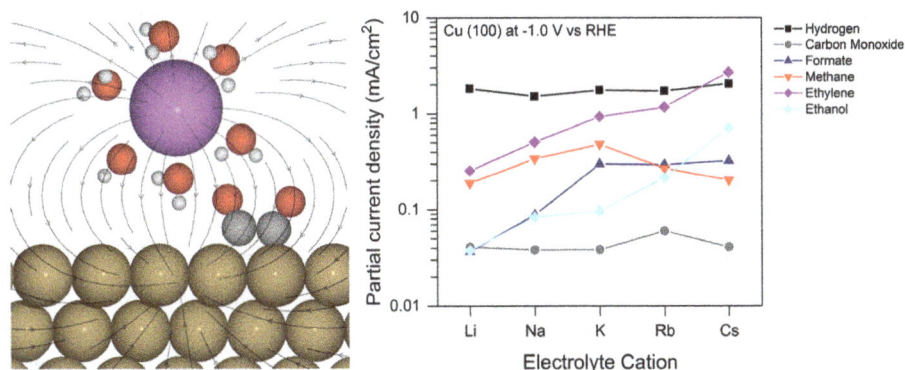

Figure 4.4: (Left) Schematic illustration of the field-stabilised $*CO$ at the surface of a Cu electrode due to the presence of a solvated cation in the outer Helmholtz layer. (Right) Partial current densities for six CO_2RR products detected on a Cu (100) single-crystal electrode in 0.1 M $XKHCO_3$, where X is one of the listed cations. Increase in cation size increases the production of formate, ethylene and ethanol but has a negligible impact on the rate of H_2, CO and methane production. Reproduced with permission from reference [249], Copyright © 2017 American Chemical Society.

Resasco et al.'s experimental data show that the larger cations increase formate, ethylene and ethanol production rate, but had minimal impact on production rates for H_2, CO or methane. This is supported by their computational data; $*H$ and $*CHO$ are not stabilised by the cations, so H_2 and methane products are unaffected, and the enhancement of $*CO$ dimerisation on stabilised surface sites means that the enhanced CO production is not seen in the product outflow. Further computational studies from

Pérez-Gallent et al. highlighted that C_2 adsorbates, specifically *OCCO and *OCCOH, were more greatly stabilised by the presence of alkaline cations than C_1 adsorbates, namely *CO and *CHO [250]. This may explain why C_{2+} product selectivity can be enhanced over C_1 by the larger alkaline cations. Rasouli et al. observed this trend under both low and high overpotential conditions at Cu electrodes, where ethylene selectivity followed $Cs^+ > K^+ > Na^+$ and methane followed the opposite trend [251].

Cation hydrolysis: Singh et al. proposed an alternate mode by which cations could enhance CO_2RR [252]. They proposed that the solvated cations within the outer Helmholtz layer preferentially undergo hydrolysis, releasing protons close to the electrode surface. In this way, the solvated cations provide a buffering capacity to the local reaction environment during CO_2RR, which helps to increase the CO_2 concentration close to the electrode surface. The increased performance with larger cations is due to the increased electrostatic interaction between the cation and the electrode surface leading to a much lower pK_a for large solvated cations close to the electrode.

It is worth noting that this mechanism is only valid for electrolytes at pH ~ 7 and for reactions where concentration of the starting material is pH dependent; the concentration of CO_2 is pH dependent because of its pH-dependent equilibrium with bicarbonate and carbonate ions. The positive impact of large cations in alkaline solutions and on the reduction of CO therefore require alternative explanations, most likely focusing on the stabilisation of bound intermediates [250].

4.2 Organic solvents

Applications of aqueous electrolytes to CO_2RR are more common than organic electrolytes, thanks in part to their low cost and high sustainability and availability. This is particularly relevant when considering up-scaling when costs and hazards associated with organic media toxicity and volatility become increasingly problematic. However, CO_2RR in organic electrolytes can often be desirable.

Many of the commonly employed organic solvents offer a comparatively high CO_2 solubility versus water. CO_2 solubility has been discussed as a key limitation in CO_2RR technologies, meaning that traditional liquid electrochemical cells can only selectively drive CO_2RR at very low reaction rates. Some organic media can offer a tenfold increase in CO_2 solubility versus water, although the situation is a little more complex when considering the total dissolved inorganic carbon for carbonate and bicarbonate electrolytes.

4.2.1 Aprotic solvents

A key advantage in moving to aprotic organic systems is the removal of the parasitic water reduction reaction. CO_2RR in an organic solvent will not produce hydrogen gas in the outflow. Organic solvents tend to have a significantly wider solvent window than water, meaning CO_2RR can be carried out at very negative potentials without the need to consider solvent degradation.

Mechanism in organic media: The absence of water in organic media means that CO_2RR cannot proceed via the same mechanism as in the aqueous case. The first stage is a single electron transfer to CO_2 to give the $CO_2^{\bullet-}$ radical anion. The next stage can proceed via one of two routes. Two $CO_2^{\bullet-}$ radicals may dimerise to give oxalate ($C_2O_4^{2-}$), or $CO_2^{\bullet-}$ and a second CO_2 molecule may disproportionate to give CO and CO_3^{2-} [258]:

$$CO_2 + e^- \rightleftharpoons CO_2^{\bullet-} \tag{4.7}$$

$$CO_2^{\bullet-} + CO_2^{\bullet-} \rightleftharpoons C_2O_4^{2-} \tag{4.8}$$

$$CO_2^{\bullet-} + CO_2 \rightleftharpoons C_2O_4^{\bullet-} \tag{4.9}$$

$$C_2O_4^{\bullet-} + e^- \rightleftharpoons CO + CO_3^{2-} \tag{4.10}$$

Which mechanistic direction is followed, and therefore which product is formed, has been reported to depend on the strength of the interaction between the catalyst and the $CO_2^{\bullet-}$ radical (Figure 4.5). If the bond is strong, then reaction with solvated CO_2 to give CO is favoured, whereas if the bond is weak the dimerisation of two $CO_2^{\bullet-}$ to oxalate is favoured [257].

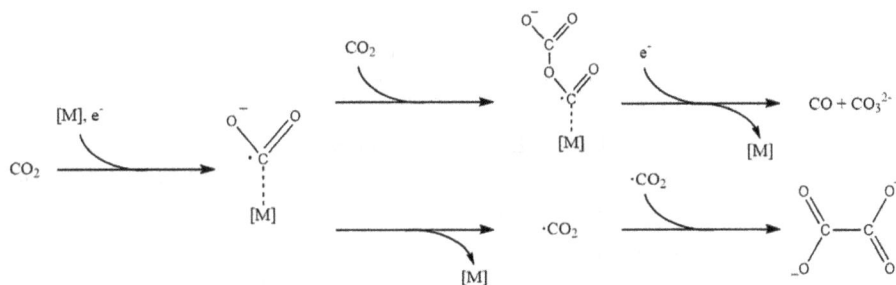

Figure 4.5: Proposed mechanism for CO and oxalate formation from CO_2RR in dry organic media. Following the first electron transfer to give the radical anion, CO_2^- may react with CO_2 to give CO via a carbon–oxygen adduct (top) or dimerise with a second radical anion to give oxalate (bottom). Reproduced with permission from reference [257], Copyright © 2019 König et al.

Role of the solvent: A number of studies have investigated how changing the employed organic solvent impacts the product distribution. Unfortunately, it is very difficult to decouple the relative impacts of the different parameters associated with organic solvents (Table 4.1), all of which seem to impact the product distribution and/or the reaction rate. In one of the earliest works, Kaiser and Heitz found that the selectivity was shifted towards oxalate in solvents with decreasing nucleophilicity. They proposed that more nucleophilic solvents coordinate to the carbon in CO_2 to hinder electron transfer [259]. Moving towards less nucleophilic solvents such as 1,2,-dichloroethane was not possible as their relative permittivity was too low to allow the electrolyte to dissociate, giving an unfavourably resistive solution [256].

Table 4.1: Solvents commonly used for CO_2RR along with their solubility, viscosity and relative permittivity. Solubility was taken as the mean of values given in references [253–255]. Viscosity, relative permittivity and donor number were taken from reference [256]. Reproduced with permission from reference [257], available open access on Creative Commons CC-BY license, Copyright © 2019 König et al.

Solvent	CO_2 solubility (mM)	Viscosity (mPa s)	Relative permittivity	Donor number (kJ mol^{-1})
Acetonitrile	314 ± 6	0.341	35.9	59.0
Dimethyl formamide	194 ± 14	0.802	36.7	111.4
Dimethylsulphoxide	131 ± 7	1.99	46.5	124.8
Hexamethyl phosphoramide	174 ± 15	3.10	29.6	162.4
Methanol	151 ± 11	0.551	32.7	79.5
Propylene carbonate	134 ± 9	2.53	66.1	63.2
Tetrahydrofuran	313 ± 40	0.460	7.6	83.7
Water	34.5 ± 4.4	0.890	78.4	138.2

Berto et al. [260, 261] investigated the CO_2RR in different organic solvents, showing that the onset potential was lowest for solvents with intermediate relative permittivity. As is seen for metal–CO_2 binding energy in the catalytic volcano plots, increase or decrease in the relative permittivity away from the optimum resulted in a decreased performance.

Unfortunately, the choice of organic solvent is not as simple as following the least negative onset potential or the greatest current density. According to Figure 4.6, the best organic solvent to investigate for CO_2RR is clearly acetonitrile. However, acetonitrile is both toxic and volatile, which raise questions over its safety for larger scale industrial applications. Similar issues surround the other listed organic solvents, since DMF is poisonous and readily hydrolyses, and DMSO is a solid below 18 °C. This led the authors to proceed with propylene carbonate as their organic

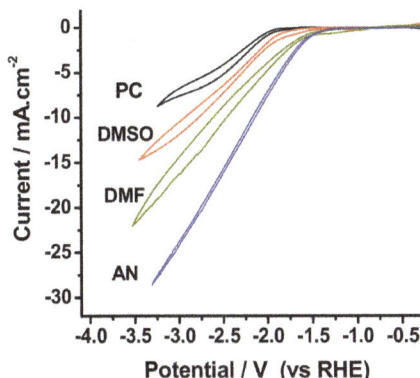

Figure 4.6: Cyclic voltammograms for CO_2 reduction at an Au foil electrode in acetonitrile (AN), dimethylformamide (DMF), dimethylsulphoxide (DMSO) and propylene carbonate (PC) with 0.1 M Bu_4NClO_4 electrolyte. Optimising the choice of organic solvent can be seen to significantly impact both the onset potential and the magnitude of the current density. Reproduced with permission from reference [261], Copyright © 2017 Elsevier Ltd.

solvent of choice; although the electrochemical performance was worst, it is non-toxic, non-volatile and has a wide potential window, making it the safest option for wide-scale use [261].

Electrolyte effects: As with aqueous systems, CO_2RR in organic solvents is greatly impacted by the choice of electrolyte. The commonly reported alkali metal salts used in aqueous systems are not sufficiently soluble in aprotic organic media, so alkylammonium perchlorates, sulfonates and tetrafluoroborates are preferred. Although the equivalent Li salts are also soluble in aprotic organic solvents, experimentally it has been shown that no CO_2RR products are detected if a Li salt is used in place of its alkylammonium equivalent [262].

It was originally suggested that the alkylammonium cation had a facilitatory role in mediating the charge transfer from the electrode to CO_2 [263]. However, recent voltammetric studies by Berto et al. have shown no impact on varying the nature of the alkylammonium cation on the rate of CO_2RR or product distribution [260]. Setterfield-Price and Dryfe provided an alternative explanation where the lack of CO_2RR in Li salts is due to a passivating Li film inhibiting CO_2 adsorption at the catalyst surface [264]. Kai et al. also used scanning electrochemical microscopy to demonstrate no ion-pairing between CO_2 and alkylammonium cations, although their data does show a cation size dependence [265]. The authors attribute this to larger cations increasing the electron tunnelling distance between the electrode and CO_2, decreasing the rate constant for the first electron transfer step.

4.2.2 Hydrated organic solvents

CO_2RR in dry, aprotic organic media leads to CO and oxalate as the dominant products, as alternative products require proton transfers that cannot be provided in the absence of water. Higher order CO_2RR products can be accessed by moving to a water-in-organic system, where water is added in low concentrations. Water therefore functions as a secondary reactant, while the bulk solvent is still organic. Organic systems are highly sensitive to the presence of water; Figueiredo et al. reported that only 46 ppm water made a measurable difference in acetonitrile [266]. The concentration must be optimised in order to provide the necessary protons for higher order CO_2RR products without being so high that the rate of the parasitic HER overwhelms the CO_2RR, particularly at higher current densities.

Mechanistic changes: Of course, a side effect of residual water in organic media is that the HER is possible as a means of producing H_2 in the product outflow. The proportion of H_2 made can be expected to be related to the water content. Similarly, the presence of water means that dissolved CO_2 may take part in the bicarbonate equilibrium that was discussed for CO_2 in aqueous systems. This is also facilitated by the water reduction reaction, which produces OH^-. In situ spectroelectrochemical studies have confirmed that OH^- can combine with CO_2 directly to give bicarbonate in wet acetonitrile [266]. These negative impacts will need to be balanced against the benefits to the water-in-organic solvent approach.

Addition of water to the organic solvent gives access to the same CO_2RR mechanism as is seen in aqueous media, where protonation steps after the initial electron transfer are now possible. This has an understandably sizeable impact on the product distribution. This protonation step means that a prevalent product in organic-water mixtures is formate. The mechanism for formate product starts with an initial protonation of the $CO_2^{\bullet-}$ radical anion. The protonated $HCO_2^{\bullet}$ can then be further reduced to give formate, either by a second electron transfer at the electrode or via a homogeneous reaction with a second $CO_2^{\bullet-}$ radical anion [267]:

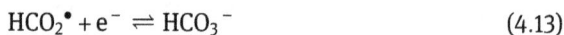

$$CO_2^{\bullet-} + H^+ \rightleftharpoons HCO_2^{\bullet} \tag{4.11}$$

$$HCO_2^{\bullet} + CO_2^{\bullet-} \rightleftharpoons HCO_3^{-} + CO_2 \tag{4.12}$$

$$HCO_2^{\bullet} + e^- \rightleftharpoons HCO_3^{-} \tag{4.13}$$

The production route for formate can be seen to be in conflict with the route to oxalate, where the facile protonation of the $CO_2^{\bullet-}$ radical prevents its dimerisation. Tomita et al. observed this at Pt electrodes, where adding water to acetonitrile gave a progressive shift in selectivity from oxalate to formate as the majority product [262]. Aljabour et al. observed the same trend for CO versus formate production at Co_3O_4

nanowires in acetonitrile. Water availability causing the protonation step competed with the CO_2–CO_2^- coupling reaction, shifting the selectivity from CO to formate [268].

It is worth noting that one recent report has been able to produce oxalate in an entirely aqueous system. Paris and Bocarsly used a Cr–Ga oxide thin-film catalyst to produce 59% oxalate at ~ 10 mA cm^{-2} [269]. This required careful refinement of the Cr:Ga ratio, electrolyte and pH, with small deviations giving a sizeable reduction in oxalate output. However, the detection of significant oxalate in an aqueous system suggests an alternative route to oxalate that does not rely on CO_2^- radical anion coupling. This certainly merits further research regarding both aqueous and water-in-organic CO_2RR systems.

Reduction rate: The addition of water to organic media also gives a more energetically favourable CO_2RR route. This is likely due to a protonation step accompanying the first electron transfer to CO_2 to give a more stable surface-bound intermediate [270]. Shi et al. demonstrated this with voltammetry at an Au electrode in propylene carbonate (Figure 4.7) [261]. Addition of 6.8% water gave a sizeable increase in current density and a shift in the CO_2RR onset to a less negative potential. Of course, considerations to this increase in total reaction rate need to be taken alongside the product selectivity. Systems designed for oxalate production would likely omit this line of research, since an increased current density due to the presence of water would most likely come with a decreased oxalate yield as $CO_2^{\bullet-}$ is so strongly disfavoured.

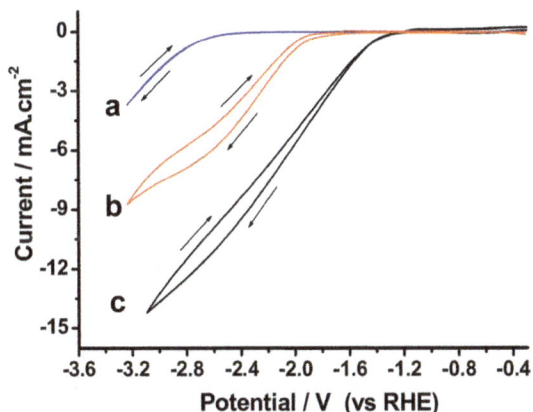

Figure 4.7: Cyclic voltammetry at an Au electrode in propylene carbonate with (a) Ar saturation, (b) CO_2 saturation and (c) CO_2 saturation with 6.8 wt% water. Addition of water gives a large increase in both the current density and onset potential. Reproduced with permission from reference [261], Copyright © 2017 Elsevier Ltd.

Higher order products: As well as facilitating the protonation of the $CO_2^{\bullet-}$ radical anion, the addition of water to organic solvents also permits further protonation steps further along the CO_2RR mechanism. This gives access to higher order reduction products that are not feasible in dry organic solvents. As the number of proton transfers is now much larger per product (8 for methane, 12 for ethylene, etc.), reported water concentrations are much higher, on the order of millimolar to molar, rather than ppm as is reported for formate.

Zhang et al. demonstrated a CuPd alloy for methane in acetonitrile with 1 M water, which produced 51% methane at ~ 6 mA cm^{-2} [209]. They proposed that the reduction of H_2O at Pd sites produced surface Pd–H, which then reduced Cu-bound CO to drive CO_2RR along the methane pathway. The methane efficiency was dependent on the Cu:Pd ratio, supporting a synergistic role between sites. Ramesha et al. produced methanol at TiO_2 electrodes in a acetonitrile/water mix [271]. They showed a strong dependence on the water concentration, where increasing water concentration from 0 to 10 mM increased the methanol concentration by a factor of 3.

Practical considerations: The access to entirely dry aprotic solvents is highly unlikely. At the small scale it is possible to do this via electrolysis, where the electrolyte is purged of water at the anode and cathode by electrochemically evolving O_2 and H_2, respectively. At the larger scale, this is unlikely to be feasible. Instead a fair degree of water removal can be achieved by storing organic solvents in molecular sieves. In this way, Subramanian et al. were able to lower the water content of their acetonitrile from 623.4 ppm to 37.6 ppm [272]. In either case, it is important to consider that any practically up-scaled CO_2RR reactor using organic solvents will have to deal with at least trace amounts of water.

4.2.3 Protic organic solvents

An alternative to the water-in-organic method is to move to a protic organic solvent. Of these, methanol is arguably the most commonly employed example, thanks to its similar pK_a to water (17.2 and 14, respectively). Combining this with the high CO_2 solubility and removal of water allows relatively high CO_2RR reaction rates even in liquid-phase H-cells, with the proportion of H_2 in the product stream greatly reduced [273].

Product distribution: The access to protons for CO_2RR means that the observed products in methanol are fairly similar to those reported in water. The dominant product is primarily defined by the choice of catalyst material, as discussed in Chapter 3 for catalysts under aqueous conditions [274–276]. One notable exception to this is where a material would normally favour HER over CO_2RR entirely. Moving to an organic solvent can allow previously inert catalysts to drive CO_2RR to a useful product. Saeki et al. produced 45% CO and 15% formate at a Pd electrode in methanol at 200 mA cm^{-2} [274]. Tomita et al. demonstrated the same effect in aprotic solvents, where Pt in acetonitrile produced 71.1% oxalic acid at 5 mA cm^{-2} [262].

Electrolyte effects: Unlike in aprotic solvents, many alkali metal salts have a similar solubility in methanol to water. As such, CO_2RR in methanol shows similar cation and anion dependencies to aqueous systems. Kaneco et al. compared CO_2 reduction at a Cu electrode methanol with NaOH electrolyte, which gave 63% methane and 17.6% ethylene [277], whereas equivalent CO_2RR in CsOH gave 4.1% methane and 32.7% ethylene [278]. The authors attributed this difference to the smaller cations providing a more hydrophilic electrode surface, favouring CO hydrogenation on the methane pathway. The use of hydroxide-based electrolytes was employed to suppress the HER, as is also done in aqueous systems [279].

4.3 Ionic liquids

Ionic liquids have a number of intrinsic properties that make them inherently attractive for CO_2RR applications. They are thermally and chemically stable with low volatility and a wide electrochemical window, making them desirable for many electrochemical systems. As with organic systems, they can be operated in the absence of water, removing challenges associated with the HER and with CO_2 entering the bicarbonate equilibrium. The scope of ionic liquids available is incredibly broad. A representative selection of available ionic liquids is given in Figure 4.8.

CO_2 solubility: A key advantage to ionic liquids for CO_2RR applications is their high CO_2 solubility. In fact, their capacity to solubilise CO_2 means that ionic liquids have been extensively studied for carbon capture and storage applications [281–283]. This presents ionic liquids as a means of unifying CO_2 capture and CO_2 reduction technologies; CO_2 captured and dissolved in an ionic liquid could then flow into a secondary reactor for conversion to a value-added product without a need for extraction or purification steps. It is worth noting that ionic liquids for CO_2 capture are not necessarily inert. CO_2 can react with imidazolium and amine ionic liquids to give the equivalent carboxylate or carbamate, respectively (Figures 4.9 – 4.10). In these cases, CO_2 reduction directly from ionic liquids would have to consider the impact of either the rate of CO_2 release or the direct reduction of the carboxylate/carbamate itself on the overall CO_2RR rate.

4.3.1 Cation effects

A number of ionic liquids have been reported to specifically enhance CO_2RR by interacting with reaction intermediates. One of the most actively researches is the imidazolium cation and its associated ionic liquids. Numerous works have shown that the CO_2RR proceeds at less negative potentials in the presence of imidazolium,

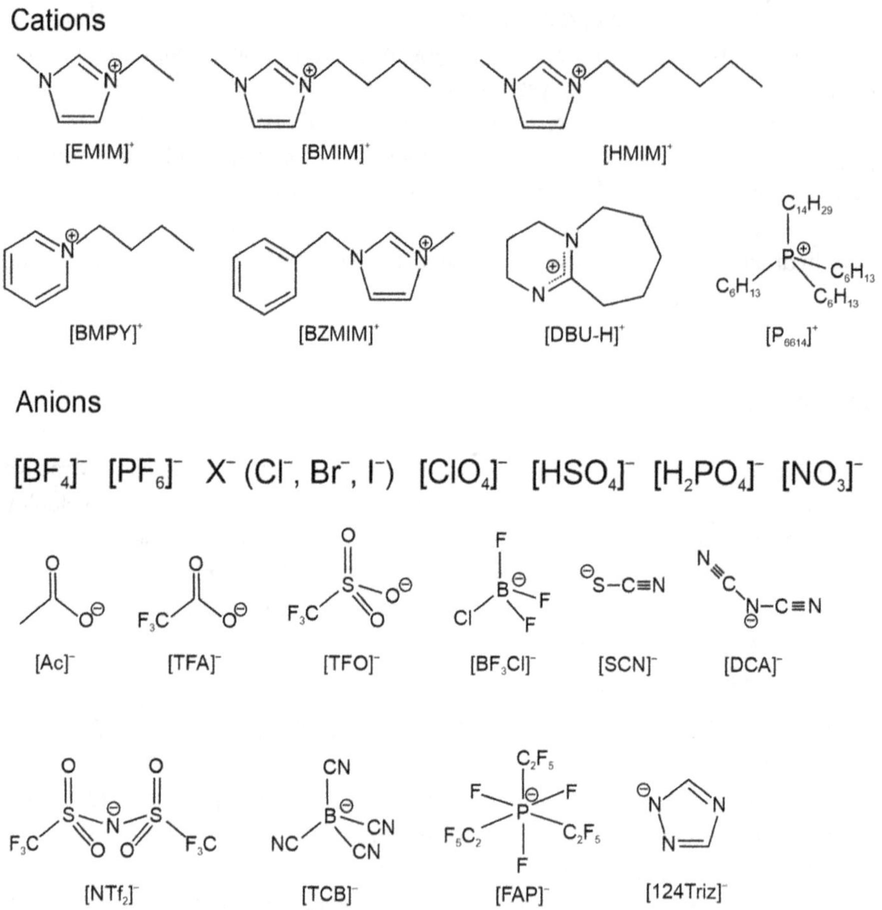

Cations

[EMIM]⁺ [BMIM]⁺ [HMIM]⁺

[BMPY]⁺ [BZMIM]⁺ [DBU-H]⁺ [P₆₆₁₄]⁺

Anions

$$[BF_4]^-\ [PF_6]^-\ X^-\ (Cl^-, Br^-, I^-)\ [ClO_4]^-\ [HSO_4]^-\ [H_2PO_4]^-\ [NO_3]^-$$

[Ac]⁻ [TFA]⁻ [TFO]⁻ [BF₃Cl]⁻ [SCN]⁻ [DCA]⁻

[NTf₂]⁻ [TCB]⁻ [FAP]⁻ [124Triz]⁻

Figure 4.8: A sample of the available cationic and anionic ionic liquid components. Adapted with permission from reference [280], Copyright © 2020 Elsevier Ltd.

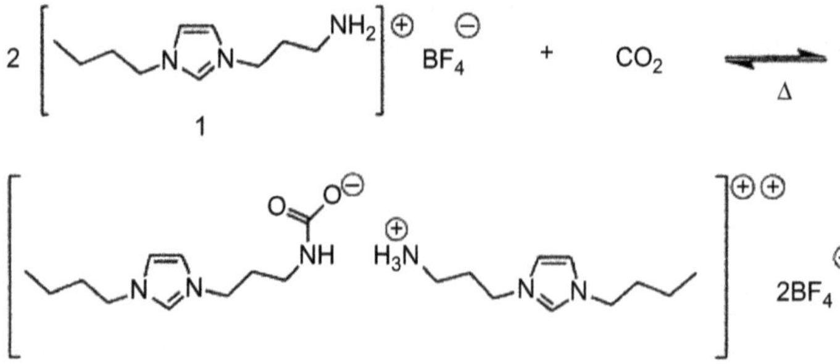

Figure 4.9: Reaction mechanism for carbamate formation through the reaction of CO_2 with an amino acid ionic liquid. Reproduced with permission from reference [284], Copyright © 2002 American Chemical Society.

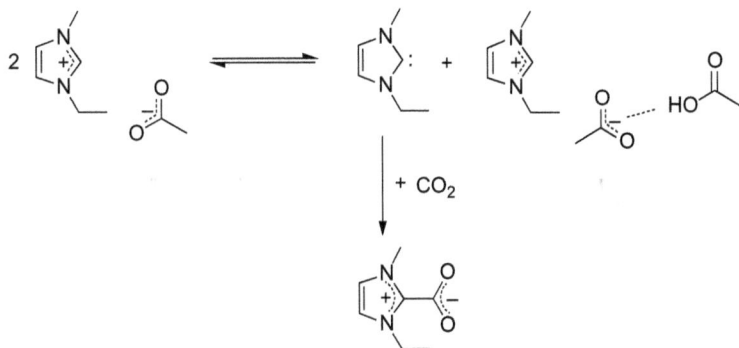

Figure 4.10: Reaction mechanism for carboxylate formation through the reaction of CO_2 with an imidazole ionic liquid. Reproduced with permission from reference [285], Copyright © 2011 Wiley-VCH Verlag GmbH & Co. KGaA, Weinheim.

indicating a more kinetic facile reaction. However, there is still some debate on the precise mechanism by which the ionic liquid has its impact.

Binding to the radical anion: Rosen et al. proposed a low-energy pathway for CO_2 reduction to CO at an Ag electrode in [EMIM]BF$_4$ [286]. The complex formed between [EMIM$^+$] and CO_2 has a much lower activation energy for the first electron transfer, allowing CO_2RR to CO to proceed with a lower overpotential. The authors also used in situ spectroscopy to demonstrate that a layer of [EMIM]$^+$ at the electrode surface can effectively hinder HER, as well as lower the activation energy for the first electron transfer [287]. The same stabilisation effect has also been proposed for CO_2RR to formate at Pt electrodes [288] and to CO at Au electrodes [289, 290].

Reduction of the ionic liquid cation: Zhao et al. suggested an alternative mechanism, where it is the imidazolium cation that is first reduced to give a neutral radial species (Figure 4.11) [291]. This then complexes with CO_2 to give a carboxylate that is more easily reduced at the electrode surface, giving CO as the main product. This mechanism highlights a source of deactivation of the ionic liquid, where carbene formation after CO loss can lead to the formation of a catalytically inactive carboxylate, although this can be prevented with sufficient proton supply within the ionic liquid.

Stabilisation through hydrogen bonding: Lau et al. suggested that, rather than C2 position being catalytically active as is shown in Figure 4.10, it is the two protons at C4 and C5 that stabilise the CO_2^- anion through hydrogen bonding (Figure 4.12) [292]. To support this, they demonstrated that attaching bulky groups to the C2 site did no impact on the catalytic enhancement, whereas replacing C4 and C5 protons with methyl groups reduced the current density and increased the required overpotential. Although the improved CO_2RR in imidazolium ionic liquids is clear, further work is still needed to resolve the origins of this in order to better design future ionic liquids for CO_2RR.

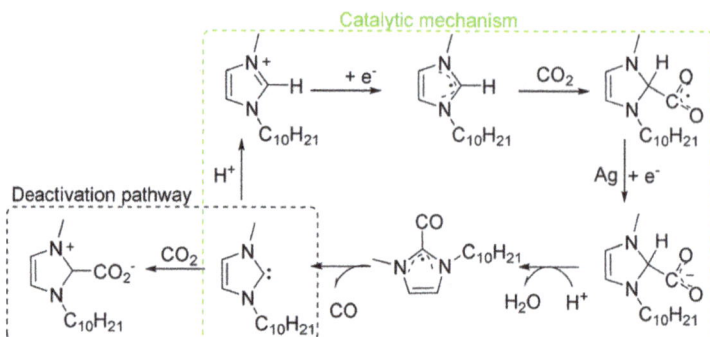

Figure 4.11: Proposed catalytic mechanism for an imidazolium ionic liquid cation ($[C_{10}MIM]^+$) on the CO_2RR at an Ag electrode. Reproduced with permission from reference [291], Copyright © 2016 American Chemical Society.

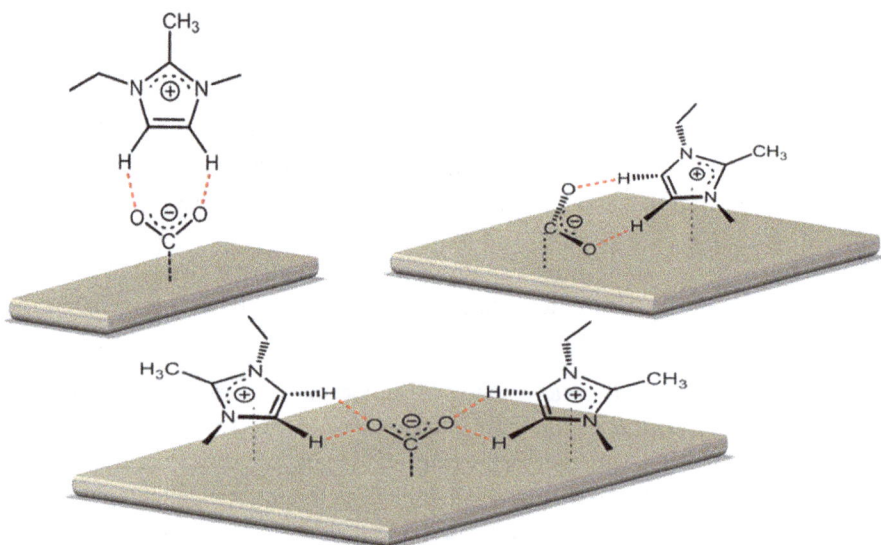

Figure 4.12: Possible binding modes between an imidazolium cation and the CO_2^- radical cation at an Ag electrode surface. Reproduced with permission from reference [292], Copyright © 2016 American Chemical Society.

Double-layer effect: Other ionic liquid cations have also been shown to facilitate CO_2RR, although equivalent mechanisms involve direct reactions with CO_2RR intermediates. Instead, these cations are proposed to enhance CO_2RR through a double-layer effect, where their positive presence at the electrode helps stabilise the negative charge on the CO_2^- radical anion. Zhao et al. demonstrated this through CO_2RR with ionic liquids with varying cation sizes [291]. As the cation was made smaller, its charge density increased and so the CO_2RR onset potential shifted towards less negative potentials, indicating a more facile reaction. Similar studies changing the anions had no significant impact.

Sun et al. attributed this effect to their observations that adding ionic liquids to an organic solvent shifted the selectivity of a Pb electrode from oxalate to CO [293]. The authors proposed that the layer of cations at the electrode surface stabilised the CO_2^- radical anion to enhance the CO pathway, while also hindering the dimerisation of CO_2^- to hinder the oxalate pathway.

4.3.2 Protonation steps in ionic liquids

Protic ionic liquids: The availability of protons in ionic liquids is not constant. Different protonated ionic liquid cations will display very different acidities based on the structure of the cation. Atifi et al. demonstrated that switching from an aprotic ($pK_a = 32$) to a protic ionic liquid ($pK_a = 24$) could shift the selectivity from ~ 85% CO to ~ 80% formate (Figure 4.13) [294]. This suggests that protic ionic liquids can facilitate hydrogenation steps in CO_2RR as well as their stabilising effect on negatively charged intermediates.

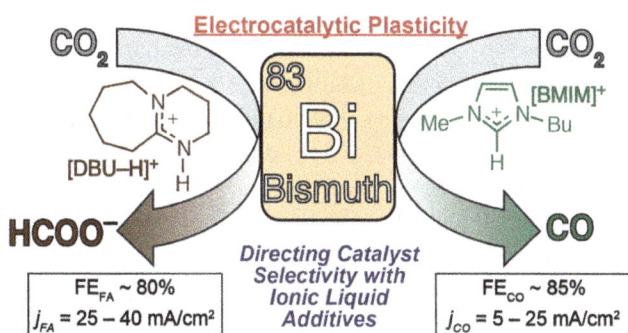

Figure 4.13: Schematic reaction mechanism change for CO_2RR in two different ionic liquids. The protic ionic liquid (brown) facilitates hydrogenation steps to give formate as the dominant product, whereas the aprotic ionic liquid facilitates the formation of CO. Reproduced with permission from reference [294], Copyright © 2018 American Chemical Society.

Water in ionic liquids: As with organic solvents, ionic liquids show limited proton availability, which leads to most CO_2RR systems producing CO with few other products detected in significant amounts. One way around this is to use ionic liquids in combination with water, either added in stoichiometric amounts or as the primary solvent, where ionic liquids are added in molar quantities to function as the electrolyte. Importantly, the presence of the ionic liquid seems to give the benefits of protonation steps to facilitate CO_2RR without lowering the charger efficiency due to

HER. Rosen et al. found this for water in [BMIM]BF$_4$, where the H$_2$ yield at an Ag electrode remained low as long as water was kept below 90% [295].

A secondary impact of adding water to ionic liquids is on the viscosity. Most ionic liquids are highly viscous, which slows the rate of mass transport to the electrode surface, giving a slower overall rate of reduction. Adding water to ionic liquids can mediate this by lowering the solution viscosity to increase the mass transport of CO$_2$ and also water to the electrode. Importantly, the diffusion coefficient increases more rapidly than the solubility of CO$_2$ decreases as water is added, indicating a compromise dosing of water is possible without negating the solubility benefits of ionic liquids [296].

It is worth noting that this is not the only means to lowering the viscosity of an ionic liquid. Ionic liquids with shorter alkyl chains tend to have lower viscosities [297]. Equally, ionic liquids could be diluted with organic solvents, either a protic or aprotic depending on the desired product. The impact of the co-solvent on the overall viscosity will depend on the polarity, with more polar solvents giving a greater reduction in viscosity.

4.3.3 Cheaper catalyst materials

Many ionic liquid studies use Ag as the cathode material. The motivation is providing inherently high CO$_2$RR activity, close to that of Au without the restrictive costs. Starting with a high activity material then allows variations in the ionic liquid to be discussed in terms of an already large selectivity and production rate. Further advancements have also come through complementary nanostructuring of Ag surfaces to give a highly active Ag surface complimented by the ionic liquid [286, 298].

The provision of a lower energy mechanistic route through the stabilisation of adsorbed CO$_2$RR intermediates opens up the possibility of using cheaper, more abundant materials in place of the more active and expensive materials such as Au and Ag. Medina-Ramos et al. probed multiple Sn, Sb, Pb and Bi electrodes for CO$_2$RR in the [BMIM]OTf ionic liquid [299]. Bi and Sn gave good ~ 77% CO at ~ 10 mA cm^{-2}. Pb also produced CO but required a much greater overpotential and was quickly passivated during CO$_2$RR. Sb produced know noteworthy CO$_2$RR. This demonstrates the improved CO$_2$RR activity in the presence of ionic liquids but only if appropriate CO$_2$– cation–electrode material interactions are chosen (Figure 4.14).

Non-transition metals: Bi has been a popular choice for CO$_2$RR to CO in ionic liquids. DiMeglio et al. demonstrated that adding ionic liquids to acetonitrile in place of the electrolyte increased the faradaic efficiency from 48% to 95% and increased the current density from 0.2 to 5.5 mA cm^{-2} [300]. Adding small quantities of water to ionic liquids allows Pb and Sn electrodes to produce fairly high amounts of formate. Zhu et al. demonstrated this for 30 wt% ionic liquids with 5 wt% water in acetonitrile. With the ratio, the choice of ionic liquid had a sizeable impact on formate yield, with [BMIM]PF$_6$ giving the best performance of 95% formate at ~ 17 mA cm^{-2} for both metals.

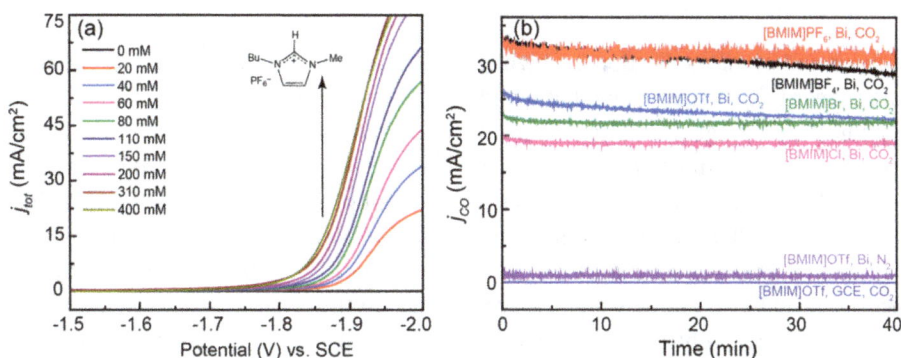

Figure 4.14: (a) CO$_2$RR current density at a Bi/C electrode in acetonitrile with varying concentrations of [BMIM]BF$_6$. Increasing the concentration gives a stepwise increase in current density. No current density is observed in the absence of ionic liquid. (b) Partial current density for CO in varying ionic liquids. Decreased performance in halide-based ionic liquids was proposed to be due to the more hygroscopic halide ionic liquids introducing water into the reactor. Reproduced with permission from reference [299], Copyright © 2014 American Chemical Society.

Activities of these materials have been improved by electrodepositing catalyst from the same ionic liquid solution under CO$_2$RR operating conditions. This in situ route to electrodeposition has been reported to provide complimentary structural features for CO$_2$RR. Ding et al. produced highly active In electrodes by electrodepositing during CO$_2$RR in [BMIM]PF$_6$ [301]. The resultant nanostructures produced up to 99% CO at ~ 15 mA cm^{-2} and were stable over 15 h. Medina-Ramos et al. used the same approach with Bi electrodeposition in [BMIM]OTf, which gave 82% CO at ~ 38 mA cm^{-2} [302]. The authors noted that the CO selectivity was decreased in the presence of halide-based ionic liquids, which are more hygroscopic and so may have introduced water to the reactor to shift the selectivity towards formate.

Carbon: One of the more surprising impacts of ionic liquids is facilitating the CO$_2$RR at carbon electrodes, which are normally considered inert for CO$_2$RR. Zhu et al. used a complex graphene oxide – multi-walled carbon nanotube composite for CO$_2$RR in [BMIM]BF$_4$ in acetonitrile [303]. The precise content of the electrolyte had a significant impact on the product yield, with a 90:10 ionic liquid:acetonitrile ratio giving 85% CO at 2.3 mA cm^{-2}.

Kumar et al. achieved a similar effect with N-doped carbon nanofibers, which produced 98% CO at ~ 3.5 mA cm^{-2} in 25 mol% [EMIM]BF$_4$ in water [304]. The authors ascribed the activity to δ^+ C sites promoted by neighbouring δ^- N sites, which facilitate the adsorption of EMIM–CO$_2$ complexes for reduction to CO. Sun et al. found that using N-doped graphene-like carbon in ionic liquids selectively produced up to 93.5% methane [305]. Although the current density is low at 1.42 mA cm^{-2}, it is interesting to see ionic liquids facilitating CO_2RR to higher order products at surfaces that are usually assumed to be inactive in aqueous media.

Copper and its alloys: Section 3.3 highlighted the unique ability of copper to produce high-order CO_2RR such as methane and ethylene with high selectivity. Ionic liquids significantly change this selectivity away from higher order products, since the larger number of proton transfers (8 for methane, 12 for ethylene) are unlikely to be feasible. However, dosing ionic liquids with a small amount of water can facilitate the hydrogenation steps needed to reach formate. Huan et al. demonstrated this at a dendritic Cu catalyst in [EMIM]BF$_4$ with 8% water (v/v) [306]. The catalyst gave 83% formic acid at 6.5 mA cm^{-2}. Interestingly, a Cu plate electrode operating under the same conditions achieved only 45% formate, indicating a strong surface structure dependence for the formate mechanism in ionic liquids. Saeki et al. used methanol as an alternative proton source in ionic liquids [263]. CO_2RR at a simple Cu wire produced 86.6% CO at 333 mA cm^{-2} in [TBA]BF$_6$. Moving to LiBF$_6$ shifted the selectivity towards methyl formate as a C_2 product, indicating that the ionic liquid cation hinders the dimerisation step.

It is still possible for Cu to access higher order CO_2RR products with ionic liquids. In these cases, ionic liquid concentrations are much lower, so the ionic liquid is essentially acting as a promotor species in an aqueous environment. Sun et al. demonstrated this at a Cu/C-doped boron nitride electrode using an electrolyte of 25 mol% [EMIM]BF$_4$ and 0.01 M LiI in water, which produced 80.3% acetic acid at 13.9 mA cm^{-2} [307]. The authors ascribed the activity to a synergy between the electrode structure and choice of electrolyte, where BN-C sites drive the first electron transfer, Cu is the site for further electron transfers and LiI facilitates C–C coupling steps.

Tuning of Cu alloys alongside an appropriate ionic liquid can provide increased methanol yields (Figure 4.15). Yang et al. produced up to 77.6% methanol at 41.5 mA cm^{-2} using a Cu$_{1.63}$Se nanocatalyst [308]. Running in 30 wt% [BMIM]PF$_6$ and 5 wt% water in acetonitrile facilitated the hydrogenation steps while keeping H$_2$ production low. Moving from PF$_6^-$ to BF$_4^-$ marked a decrease in current density and faradaic efficiency, which was ascribed to more favourable PF$_6$–CO_2 interactions enhancing CO_2 solubility [309]. Similarly, Lu et al. produced a Pd$_{83}$Cu$_{17}$ aerogel that provided 80% methanol at 31.8 mA cm^{-2} [310]. This catalyst worked with substantially higher water content, using 25 mol% [BMIM]BF$_4$ and 75 mol% water. Adding water to dry [BMIM]BF$_4$ increased both the current density and methanol selectivity up to this percentage, after which the current density continued to increase but along with a loss in selectivity, presumably due to accelerated hydrogen evolution.

Metal chalcogenides: As well as variances to the choice of ionic liquid, research into ionic liquids for CO_2RR applications revealed metal chalcogenides as viable catalysts for CO production. MoS$_2$ gives highly active CO_2 reduction, thanks to the presence of

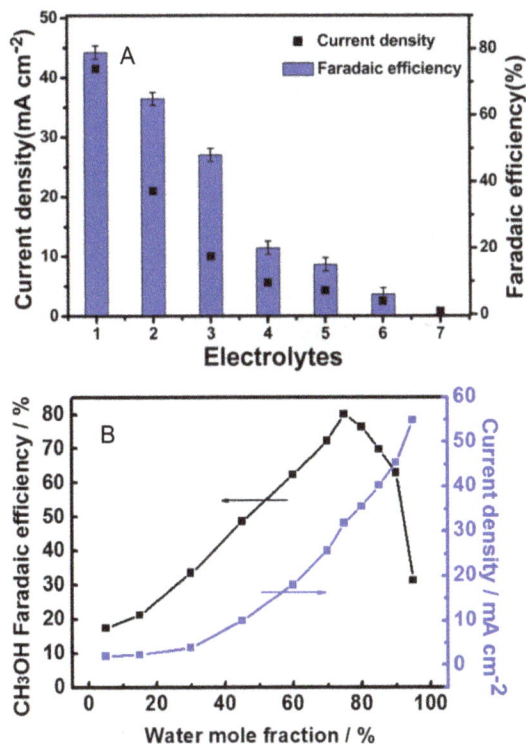

Figure 4.15: (A) Current density and faradaic efficiency for a $Cu_{1.63}Se$ electrode for CO_2RR to methanol using [BMIM]X ionic liquid with 5 wt% water in acetonitrile, where X is (1) PF_6, (2) TF_2N, (3) BF_4, (4) OAc, (5) NO_3 and (6) ClO_4. (7) was recorded in [TEA]PF_6. Reproduced with permission from reference [308], Copyright © 2019, Yang et al. (B) Current density and faradaic efficiency for methanol at a $Pd_{83}Cu_{17}$ aerogel in [BMIM]BF_4 as the mole fraction of water is increased. Additional water increases both the selectivity and current density until 75 mol%, when additional water increases current density but decreases the methanol selectivity. Reproduced with permission from reference [310], Copyright © 2018 Wiley-VCH Verlag GmbH & Co. KGaA, Weinheim.

Mo-terminated edges. These can be selectively engineered by synthesising vertically aligned MoS_2, as demonstrated by Asadi et al. to give 98% CO at −65 mA cm^{-2} in 4 mol % [EMIM]BF_4 [311]. As well as the activity of the Mo sites, the overall CO_2RR was further enhanced by the ionic liquid forming a complex with aqueous CO_2, which gave the added benefit of hindering the rate of the parasitic water reduction reaction.

MoS_2 has received growing interest since computational studies have indicated that transition metal–doped MoS_2 can break the CO_2RR scaling relationship between *CO and *CHO [312]. *CO at the doped MoS_2 surface preferentially binds to the metallic dopant, whereas other intermediates (*COOH, *CHO and *COH) preferentially bind at the sulphur site [313]. The end result is a weaker binding of *CO, allowing the CO product to be released to recover the catalytic site without disfavouring the other surface-bound intermediates of the CO formation mechanism.

Unsurprisingly, the overall performance is highly dependent on both the choice and scale of the catalyst dopant and the choice and concentration of the ionic liquid. Lv et al. improved the conductivity of MoS_2 catalysts through N-doping, giving 90.2% selectivity at 36.2 mA cm^{-2} [314]. Abbasi et al. demonstrated a highly active Nb-doped MoS_2 in 50% aqueous [EMIM]BF$_4$, which gave 82% CO at 237 mA cm^{-2} [315]. Selective doping has also allowed metal chalcogenides to produce higher order CO_2RR products. Sun et al. doped MoS_2 with Bi to give 71% methanol at 12.5 mA cm^{-2} [316]. The Bi sites provide the initial reduction to CO, which is then hydrogenated via the Mo site.

5 Reactor engineering

5.1 Electrode architectures

5.1.1 Foil and mesh electrodes

The scope of electrode architectures available to CO_2RR researchers is very broad. As a generalisation, many decisions regarding the choice of electrode architecture are based on a compromise between the activity of the electrode and the complexity of the design; producing electrodes to encourage high activity or rapid turnover rates inherently increases the complexity of the system.

Probably the simplest architecture that can be employed is the foil electrode. This is simply a piece of foil made from the target catalyst material that is immersed in a CO_2-saturated electrolyte. The key motivation for foil electrodes is the simplicity; geometry can be well defined by cutting the foil to shape or isolating an area with a non-conductive sealant. Importantly, foil electrodes are relatively straightforward to modify with higher surface area materials. A number of different techniques have been successfully employed, including templated electrodeposition, sputtering and thermally annealing [174, 177, 184]. This provides a means to rapidly assess a broad range of different structures using relatively cheap and readily available materials.

Moving from foil to mesh electrodes offers a lot of the same advantages, but with a considerably increased surface area. The uniformly conductive surface means that the same deposition techniques can be applied in virtually all cases, although line of sight may need to be considered for sputtering materials onto meshes. The inherently larger surface area can be useful in assessing these materials as measurable concentrations of CO_2RR products can be detected over shorter timescales.

5.1.2 Nanostructured materials

Many of the nanostructuring techniques discussed in Chapter 3 are CO_2RR product specific, creating crystal facets or active sites that are specifically designed to favour certain reaction intermediate or mechanistic step. However, it is worth mentioning some structural changes that have measurable impact on the CO_2RR regardless of material or target product.

Increased surface area: Nanostructuring increases the electrochemically active area of electrode materials without impacting the geometric area (Figure 5.1). This greatly increases the number of active sites and therefore the turnover rate, while allowing electrode components to remain small and compact. As well as the size benefits, the larger active area per unit mass of deposited material is highly important to minimise

https://doi.org/10.1515/9781501522239-005

the costs of more expensive CO_2RR catalysts such as Ag and Au; an up-scaled Au electrode is understandably more feasible if it is made up of a thin layer of Au particles on a Ni mesh, rather than using a pure Au mesh as the electrode scaffold. Many works on CO formation at Au nanostructures focus on engineering highly active Au particles, giving the largest activity for the lowest Au mass [101, 102].

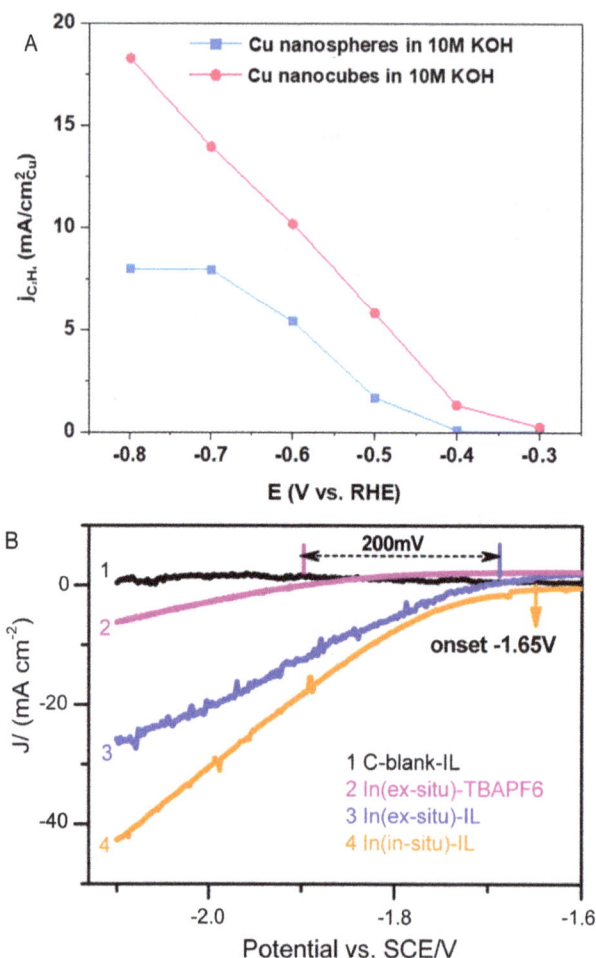

Figure 5.1: (a) Ethylene current density ($j_{C_2H_4}$) recorded at Cu nanocube and Cu nanosphere catalysts. Nanocubes preferentially exhibit the (100) facet, which favours ethylene formation over other CO_2RR products. Reproduced with permission from reference [317], Copyright © 2019 American Chemical Society. (b) Current density (j) at a glassy carbon electrode (black) loaded with ex situ electrodeposited In from $TBAPF_6$ electrolyte (pink), ex situ from ionic liquid (blue) and in situ with ionic liquid (yellow). In situ deposition during CO_2 reduction in ionic liquids created an In surface that was highly active for the CO_2RR for CO. Reproduced with permission from reference [301], Copyright © 2016, American Chemical Society.

Active crystal facets: The surface dependence of CO_2RR makes the reaction rate and selectivity highly sensitive to the array of crystal facets at the electrode surface. Templated deposition can facilitate this through the targeted production of nanostructures with the desired crystal faces. One of the common approaches is using capping agents during particle formation, as is often done for producing Cu nanocubes that are abundant with the Cu(100) facet [317]. Some recent works have shown that carrying out the catalyst deposition while CO_2 is simultaneously reduced can produce facets that are specifically active for CO_2RR. Interestingly, this has been successfully achieved for multiple materials for multiple products, including Cu for ethylene [177], and In [301] and Bi [302] for CO.

Increased retention time: The nature of the nanostructure can impact whether species close to the electrode surface stay at the electrode–electrolyte interface or diffuse away into the bulk solution. A key consideration here is the CO species. When CO is released, it may enter the gas outflow as a product species, or it may re-adsorb onto the electrode for further reduction. CO reduction proceeds via the same mechanism as CO_2 reduction, so the recapture of CO is a viable approach to increasing the yields of higher order CO_2RR products such as methane and ethylene [201].

Nanostructures with highly porous surfaces can directly impact this. Yang et al. probed the impact of nanopore depth on C_2 product selectivity [318]. They showed that an increase in pore depth increased the retention time of key intermediates, shifting the selectivity from 38% ethylene to 46% ethane. Zhuang et al. used a similar approach in creating nanoparticles with cavities designed to trap CO to drive further coupling reactions, which resulted in one of the largest recorded faradaic efficiencies for propanol production by CO_2RR (21% at 7.8 mA cm^{-2}). Ma et al. highlighted that these structures can also impact the surface pH using nanowire arrays of varying lengths [319]. Increasing the length of the nanowires increased the chances of CO dimerisation and also increased the local pH, with the longest wires showing increased yields of ethylene, ethanol and propanol.

Mass transport and gas evolution: The different types of accessible nanostructures extend far beyond nanoparticles and nanopores. High-curvature structure such as nanodendrites and nanoneedles can provide high surface areas with a large density of highly active catalytic sites (Figure 5.2) [320]. They have additional benefit of promoting nucleation of smaller gas bubbles, which provides bubble-induced mass transport to increase the CO_2RR rate [321].

Field-induced concentration effects: Catalyst surface features that are sharp on the nanoscale display high local electric fields at the tips of the nanoscale features. The concentrated electric field can significantly lower the CO_2RR activation energy barrier at these sites. Safaei et al. demonstrated this for Au nanostructures, where systematic passivation of electroplated Au gave a high density of nanosharp features, giving 95% CO at 15 mA cm^{-2} [322].

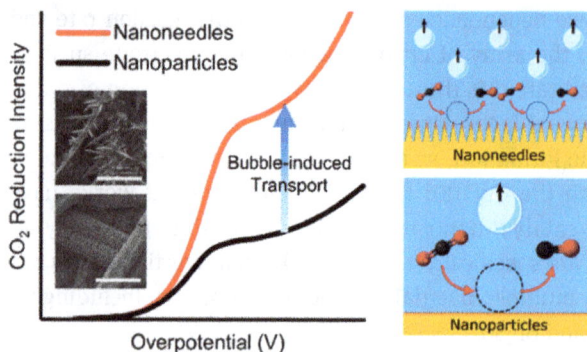

Figure 5.2: (Left) Schematic CO₂RR intensity for nanoneedle (red) and nanoparticle (black) Cu catalyst. (Right) Schematic diagrams for both catalyst surfaces, showing increased mass transport away from the nanoneedle surface due to bubble evolution. Bubble-induced mass transport from nanoneedles results in an increased CO_2RR reaction rate. Reproduced with permission from reference [321], Copyright © 2017, American Chemical Society.

The locally high electric fields on nanosharp catalyst features also concentrated cations around these sites. The choice electrolyte cation has a significant impact on product selectivity, as discussed in Section 4.1. Providing a locally enhanced concentration can therefore stabilise adsorbed CO_2 to give a locally high CO_2 concentration at the electrode surface and can also stabilise the key CO_2RR intermediates *CO₂ and *COOH [323]. Liu et al. demonstrated that nanoneedles of Au and Pd surpassed equivalent nanorods, nanoparticles and oxide-derived surfaces for the production of CO and formate, respectively [324].

5.1.3 Gas diffusion electrodes

Gas diffusion electrodes (GDEs) are ubiquitous in CO_2RR research. The basic structure is a catalyst material deposited on a porous, hydrophobic support. The GDE is then fixed into a cell so that the catalyst is electrolyte-facing and the back is gas-facing. Gas flows through the back of the GDE towards the electrolyte phase. CO_2 then dissolves in electrolyte in close proximity to the catalyst material, where it can be rapidly reduced to the target product.

GDEs and mass transport: Employing GDEs for CO_2RR keeps CO_2 mass transport in the gas phase for as long as possible (Figure 5.3). This is inherently much faster than the mass transport of dissolved CO_2 in liquid solutions, so much higher current densities are accessible [53]. This same impact can negate the low solubility of CO_2 in aqueous media. As CO_2 reacts soon after it dissolves at the gas|electrolyte interface, a high reaction rate can be attained without the need to push the solvent solubility limits [325].

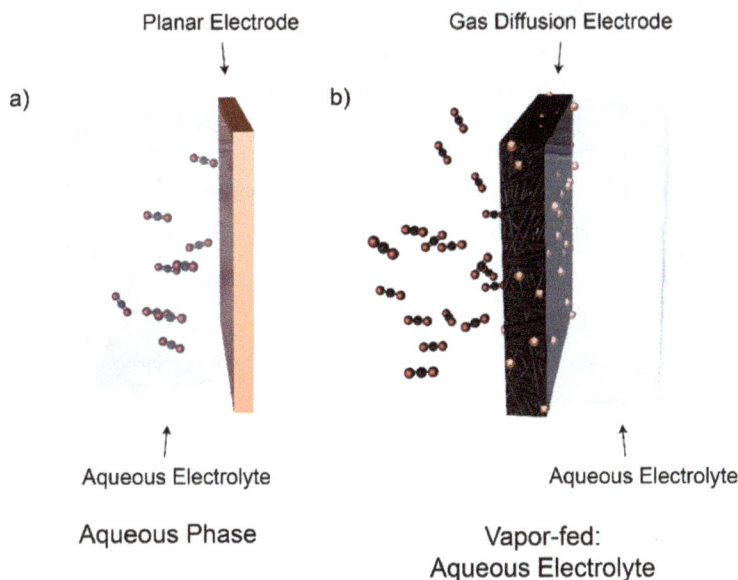

Figure 5.3: Schematic diagrams for (a) a planar electrode in an aqueous electrolyte and (b) a porous gas diffusion electrode separating a gas phase from an aqueous electrolyte. Adapted with permission from reference [325], Copyright © 2019, American Chemical Society.

Carbon GDE scaffolds: The majority of GDEs seen in the literature are porous carbonaceous materials. These have the advantage of being relatively cheap and easy to produce, as well as chemically stable at the cathodic potentials used for CO_2RR. A further advantage is their high conductivity. This permits the use of a conductive flow field to contact the back face of the GDE so that the same component both supplies CO_2 to the GDE and collects the electrical current, reducing the overall complexity of the final reactor. Carbon papers are the simplest and most commonly used structures, although moving to materials such as carbon cloth allows the same concepts to be applied to flexible electrodes and devices [132].

Hydrophobicity and GDE flooding: In order to function as GDEs, the carbon materials must be sufficiently hydrophobic to separate the gas and liquid phases. If the electrolyte enters the GDE structure it will fill the pores. CO_2 would then have to dissolve in the electrolyte within the GDE structure and diffuse in the liquid phase before reaching the catalyst layer. This phenomenon, known as GDE flooding, negates the benefits of GDEs, since rapid gas-phase mass transport is no longer possible. A further challenge for CO_2RR is that the conductive carbon support is highly active for water reduction, so a flooded GDE will produce large amounts of H_2 while also impeding CO_2 flow to the catalyst, giving a two-pronged decrease to the CO_2RR efficiency.

To combat this, hydrophobic components are introduced into the GDE structure (Figure 5.4). Commercial GDEs come loaded with hydrophobic polymers such as polytetrafluoroethylene (PTFE) or Teflon to prevent this, although the leading carbon GDEs experience flooding during extended operations so further improvements are clearly needed [326]. To this end, a number of groups have created GDEs with surface coatings that are designed to increase hydrophobicity, including oxidised carbon nanotubes [327], PTFE [328] fluoroalkylsilane [188] and dimethyl silicon oil [329].

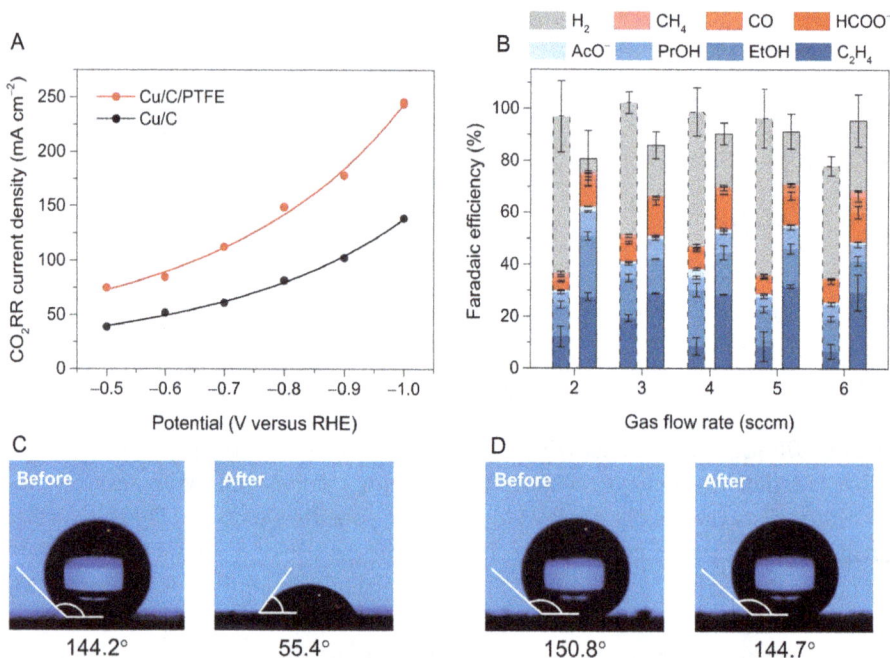

Figure 5.4: (a) Current density for CO_2RR at a Cu/C GDE with (red) and without (black) an additional hydrophobic PTFE component at the GDE surface. (b) Product distribution with (solid lines) and without (dashed lines) the hydrophobic PTFE component recorded at −1 V versus RHE with varying CO_2 flow rates. Error bars represent one standard deviation across three repeat measurements. Contact angle measurements before and after a 2 h electrolysis period for Cu/C GDEs with (d) and without (c) the added PTFE at the GDE surface. Adapted with permission from reference [328], available open access with Creative Commons CC-BY license, Copyright © 2021 Xing et al.

Other groups took a different approach and removed the conductive carbon support entirely, replacing the carbon with a porous hydrophobic polymer [225]. The resultant GDE showed impressive stability, allowing continuous CO_2RR to ethylene with ~ 70% efficiency over 150 h, with a moderate decrease in current efficiency from 100 to 75 mA cm^{-2}. Li et al. took the same approach with Au particles for CO_2RR to CO, producing up to 92% CO at 25.5 mA cm^{-2} at low overpotentials [330].

Microporous layer: Many GDEs feature a microporous layer at their surface. This is usually made up of carbon nanoparticles in order to provide a rough, high surface area interface upon which the catalyst can be deposited [331]. As well as providing a large surface area, the microporous layer can improve the conductivity of the GDE surface to reduce ohmic losses. The porous structure can encourage wetting via capillary forces, drawing electrolyte into the microporous layer [332]. The degree of wetting will depend on the hydrophobicity of the carbon material employed and can be manipulated by dosing the layer with hydrophobic components, as previously discussed. The degree of wetting should be a compromise, since water should be kept out to the pores to facilitate gas flow, but water is still needed to supply protons for the CO_2RR mechanism.

Self-supported GDEs: An interesting alternative to traditional GDEs is to build the GDE scaffold using the same catalyst material as it is applied to the electrode surface. This means that all conductive elements of the GDE are comprised of materials designed for the target reaction, that is, CO_2RR, whereas carbon GDEs almost exclusively drive the HER when wetted. This concept was demonstrated by Zhang et al., who deposited Cu_xO nanowires onto a Cu mesh to produce GDEs that gave ~ 29% ethylene at up to 300 mA cm^{-2} [333].

It is worth noting that metallic mesh bases for GDEs are often poorly hydrophobic. Coating with hydrophobic polymers such as PTFE is often necessary alongside the addition of the catalytic layer, although some catalyst surfaces can offer a fair level of hydrophobicity depending on the structure and degree of oxidation [334]. It is also important to consider the costs of the materials employed. Self-supported Au or Ag electrodes are unlikely to be feasible, but cheaper and more abundant Cu, Ni or Zn GDEs are of continued interest.

The triphasic interface: A common theme in CO_2RR research at GDEs is discussed regarding the triphasic interface. This is defined as the point where gas-phase CO_2 reacts with liquid-phase solvent at its interface with a solid-phase catalyst. This is a key discussion in the merits of GDEs over liquid-phase CO_2RR cells, since CO_2 reacting in the gas phase means that CO_2 mass transport is continuing the gas phase rather than occurring in the much slower solution phase.

Although mentioned in virtually all modern GDE discussions on the CO_2RR, the presence of a truly triphasic interface is not guaranteed. Wetting of the microporous layer or the GDE structure can result in CO_2 having to dissolve in the electrolyte and then diffuse to the catalyst (Figure 5.5). This can be thought of as a double-phase boundary, since aqueous dissolved CO_2 is reacting at the solid catalyst interface. Nesbitt et al. proposed three distinct situations that may arise in a wetted GDE: (a) the catalyst particles are completely immersed in electrolyte, giving a double-phase boundary; (b) catalyst particles are covered in a thin film of electrolyte with gas voids between them, giving a double-phase boundary; and (c) electrolyte partially wets the catalyst particles, giving multiple triphasic boundaries throughout the structure [335]. Of these, the authors proposed that CO_2RR properties of GDLs are defined by the double- rather than the triple-phase boundary.

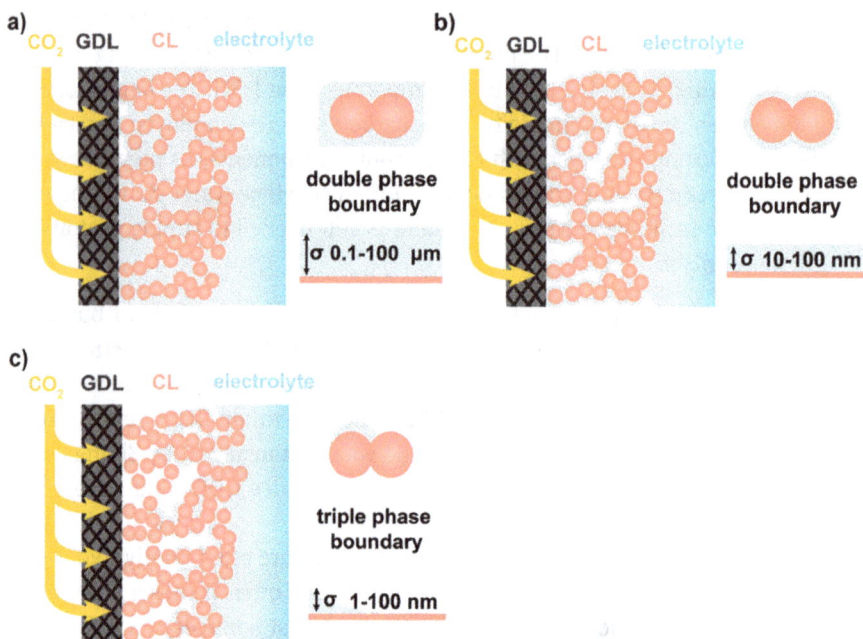

Figure 5.5: Three proposed scenarios for the presence of double- or triple-phase boundaries within the GDE structure based on the level of wetting of the catalyst layer (CL). (a) CL is fully wetted with no gas-filled voids. (b) CL is fully wetted giving a thin film of electrolyte over the particles with gas-filled voids present. (c) CL is partially wetted, giving many triple-phase boundaries. Reproduced with permission from reference [335], Copyright © 2020 American Chemical Society.

Whether you consider a double- or triple-phase interface has clear ramifications on the way in which GDE design is approached. A triple-phase interface will require a highly hydrophobic catalyst layer to ensure the catalyst is not fully wetted, whereas a double-phase interface requires a more moderate level of hydrophobicity as well as a larger pore size so that gas voids can facilitate CO_2 mass transport to reach each individual wetted particle. The key difference is arguably the dependence on the electrolyte: in the double-phase model, CO_2 must dissolve first in the electrolyte before binding to the catalyst, so factors such as solubility and the bicarbonate equilibrium still need to be considered. In both cases, it is necessary to consider the movement of water within the catalyst layer, since CO_2RR requires a continued water supply to provide protons for the reaction mechanism [37]. A commonly employed compromise is to use a hydrophilic catalyst layer alongside a hydrophobic GDE scaffold in order to facilitate CO_2 flow while providing sufficient water transport for CO_2RR [336].

5.1.4 Polymer coatings and additives

The rate of CO_2RR and its selectivity to specific products is guided by the reaction environment at the electrode surface, including the availability of CO_2, the adsorption and stability of CO_2RR intermediates and the pH at the electrode–electrolyte interface. Surface modifications to the catalyst layer can have a targeted impact on one of many of these parameters in order to drive faster reaction rates or improved product selectivity.

Ionomer layers: Nanoparticle catalysts immobilised onto electrode surfaces are typically used in combination with an ionomer layer such as Nafion. Typically, this is done to help to immobilise the catalyst particles onto the electrode as well as helping to prevent deactivation by delamination or particle agglomeration and enhance ionic conductivity [337]. A number of works have proton-conducting ionomers to enhance the proton supply for CO_2RR, which has especially helped to increase the production of higher order CO_2RR products.

Interestingly, Nafion has been used to enhance both methane and ethylene performance at Cu electrodes despite their distinct reaction pathways requiring separate means of enhancement. Pan and Barile optimised the thickness of a Nafion layer on top of a Cu foil catalyst to facilitate proton transport and hinder CO release for further reduction, giving up to 88% methane at ~ 1 mA cm^{-2} [338]. De Arquer et al. combined a thin Nafion layer with a Cu-coated PTFE fibres to give ~ 60% ethylene at current densities >1.5 A cm^{-2} [339]. The structure was designed with both hydrophilic and hydrophobic components to hinder electrode flood while maintaining ionic conductivity. Other works have also stated the importance of maintaining bulk hydrophobicity while using ionomers to facilitate proton supply for CO_2RR [336]. Clearly, the presence of the ionomer can have a different role on product selectivity depending on the material choice and electrode structure.

CO_2 availability: With CO_2 solubility providing such a challenge for CO_2RR systems, a number of groups have used surface coatings and modifications to increase the locally available CO_2 concentration. Wakerley et al. used hydrophobic Cu dendrites to trap CO_2 in a liquid cell [340]. The hydrophobic interface increased the ethylene selectivity from 9% to 56% and ethanol selectivity from 4% to 17% versus the equivalent hydrophilic surface. Perry et al. used microporous polymer coatings at Cu nanoparticles in order to trap CO_2 at the electrode surface (Figure 5.6) [341]. The authors proposed that the porous layer could also trap released CO to encourage further reduction to ethylene.

Intermediate stability: Stabilisation of key CO_2RR intermediates can lower the activation energy for the reaction. Chapter 4 demonstrated that variations in electrolyte components can achieve this by interacting with charged or radical intermediates. Other groups have achieved this same effect by tethering stabilising species onto

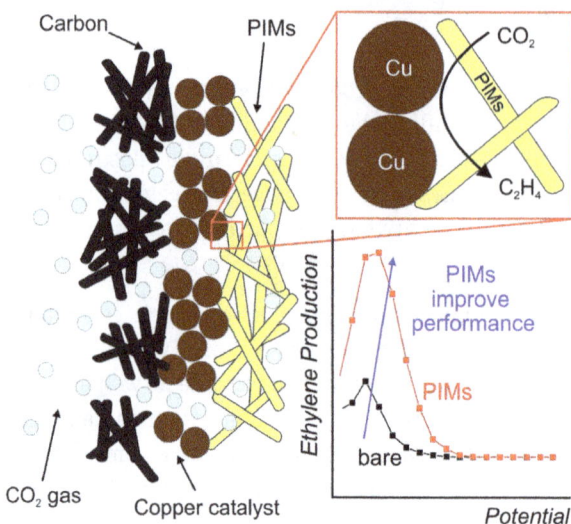

Figure 5.6: Schematic design for a Cu-coated GDE with polymers with intrinsic microporosity (PIMs). The microporous PIMs trap CO_2 and CO at the electrode surface to facilitate CO_2RR to ethylene. Reproduced with permission from reference [341], Copyright © 2020 Elsevier Ltd.

the electrode surface to work alongside the catalyst. Xie et al. modified Cu with glycine to increase the selectivity towards ethylene and ethane [342]. DFT calculations suggested that the $-NH_4^+$ moiety stabilised the *CHO intermediate. Similarly, Kim et al. used cysteine on Ag nanoparticles to stabilise the *COOH intermediate to enhance CO production [343].

Alternative mechanism: Many surface additives build upon pyridine and pyridinium moieties. Han et al. demonstrated that the choice of additive has a significant impact on the product selectivity [344]. A number of N-aryl pyridinium films shifted the selectivity from C_1 to C_{2+}, with film made up of N-tolyl pyridinium gave 78.2% C_{2+}, whereas an aryl-substituted imidazolium film produced mostly H_2. Pyridine and pyridine-derived additives have increased methanol yields at a number of different metal surfaces, including Pt, Pd and C that are normally viewed as inactive for methanol production by CO_2RR [345–347]. Yang et al. demonstrated that the N site on the pyridine is essential for methanol production at Pd, and proposed that the pyridine acted as a redox mediator to drive CO_2 reduction [345]. Giesbrecht and Herbert demonstrated that a similar effect can be achieved through the addition of N-heterocycles or dihydropyridines to working solutions (Figure 5.7) [348]. They proposed a mechanism where surface-bound organics facilitated hydride transfer to CO_2 to favour formate or methanol production.

Local proton source: The need for multiple protonation steps in CO_2RR, particularly for higher order CO_2RR products, has prompted several groups to modify catalyst surfaces with species designed to supply protons as part of the CO_2RR mechanism.

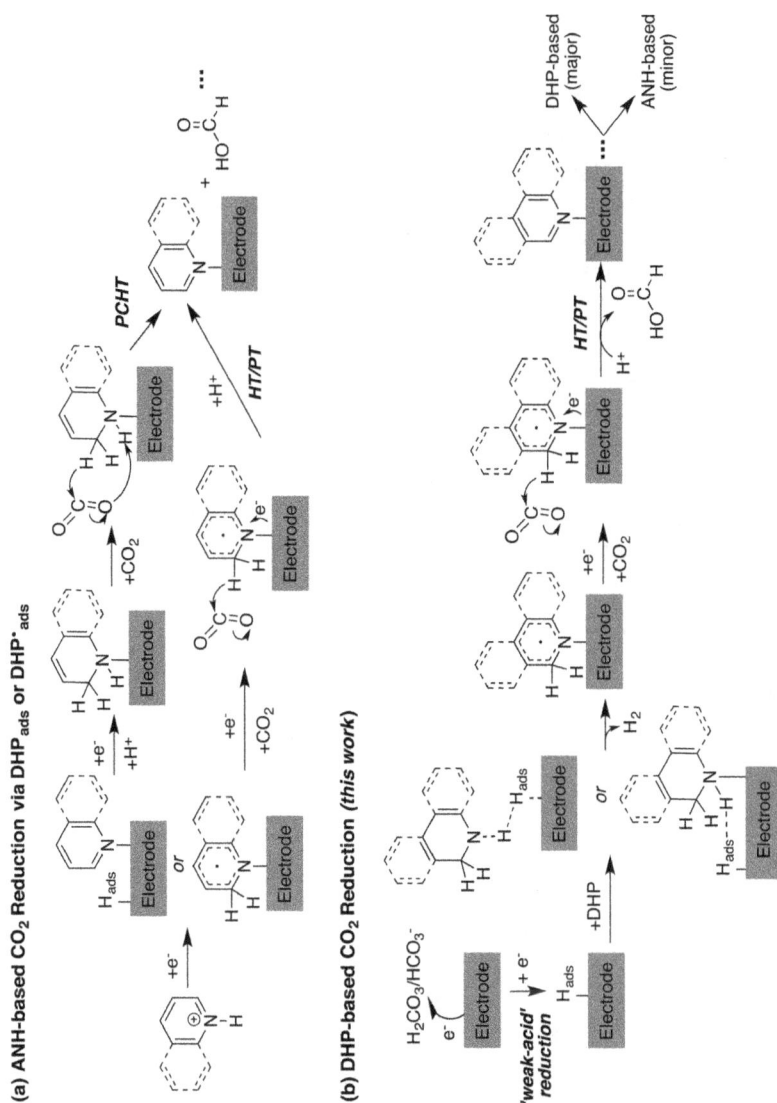

Figure 5.7: Proposed mechanisms for CO_2RR via aromatic *N*-heterocycle (ANH) and dihydropyridine (BHP) additives. Adsorption of the organic additive and subsequent electron transfers facilitates CO_2RR to formate and methanol via proton transfer (PT), hydride transfer (HT) and proton-coupled hydride transfer (PCHT). Part a was adapted from reference [349] by the authors of reference [348]. Adapted with permission from reference [348], Copyright © 2017 American Chemical Society.

A number of options have been investigated, including various phenolic [350–352] and carboxylic acid [353] compounds. Zhou et al. demonstrated that functionalising Fe with proton donor groups facilitates CO_2RR to CO even in anhydrous media, with the carboxyl acid groups providing the necessary proton transfers [353]. Calculations from Costentin et al. suggested that eight phenolic protons bound to a molecular Fe catalyst gave the equivalent increase in CO_2RR turnover frequency of 150 M phenol in solution, indicating that localised proton supply is an effective means to overcome solution limitations [350].

5.2 Reactor designs

5.2.1 Divided and undivided reactors

One of the first strategic decisions in reactor design is whether to operate in a divided or an undivided cell (Figure 5.8). Divided cells separate the anode and cathode along

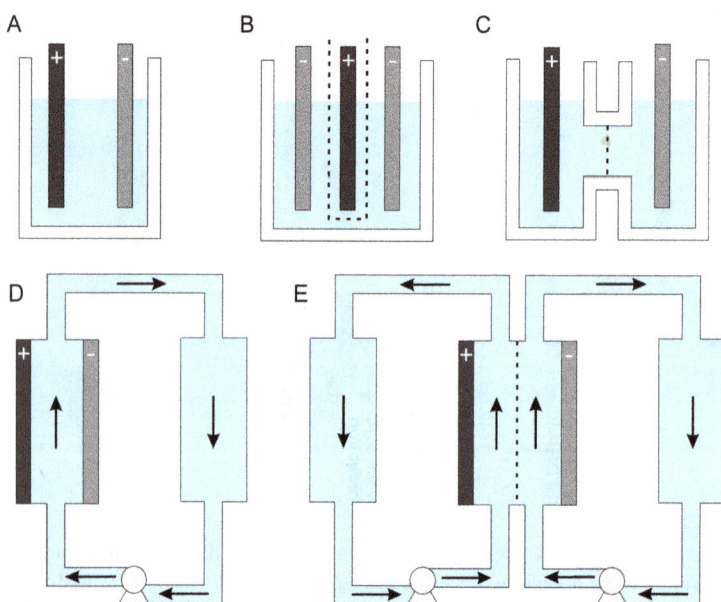

Figure 5.8: Schematic designs for several commonly used lab-scale reactor designs. (A) A simple, undivided beaker cell. (B) A beaker cell with a divided anode contained within a porous chamber. (C) A divided H-cell with a membrane separator. (D) An undivided flow cell with an electrolyte reservoir and a circulating pump. (E) A divided flow cell with separated anolyte and catholyte reservoirs on either side of a membrane separator. Solid lines indicate the direction of solution flow. Dashed line represents a separator material, which may be a porous or ion-conducting membrane. Reproduced with permission from reference [53], available open access via Creative Commons CC-BY license, Copyright © 2020 Perry et al.

with the anolyte and catholyte in two different compartments using an ion-conducting membrane or porous separator material. Undivided cells operate the anode and cathode in the same compartment with the same electrolyte volume.

Undivided cells: Single compartment designs have the advantage of simplistic design and assembly, which, as well as being easier to operated, can reduce the chances of leakage by removing the need for connecting compartments. They can also be cheaper as cell designs are easier to fabricate in house on a variety of scales, and the single compartment set-up removes the costs associated with porous separators, gaskets and other fitting required to connect the two compartments. As such, undivided cells are excellent starting points for quick verification of catalyst materials or parametric analysis of electrolyte components, especially on smaller scales.

The primary challenge associated with single compartment cells comes when species that are electrochemically generated at one electrode interact with the other. This is particularly prevalent with liquid-phase product generation during CO_2RR. For example, formate generated at the cathode could diffuse towards the positively charged anode and be re-oxidised back into CO_2. It is also important to consider pH gradients that may emerge during continued operations. CO_2RR at the cathode is usually balanced by the oxygen evolution reaction (OER) at the anode. The continued OER following eq. (1.22) causes the solution environment around the anode to become more acidic as more protons continue to be generated. As the timescale increases, this low pH diffusion field will approach the cathode, which will impact the rate of CO_2RR and also the stability of pH-sensitive components. In order to progress to greater stability over long-term operations and to protect electrochemically generated liquid products, it is often necessary to move to the more complex divided cell designs.

Divided cells: Divided cells can prevent the mixing of electrochemically generated products from either anode or cathode by physically separating them with an ion-conductive or porous membrane. The membrane must be chosen to provide reasonable conductivity, either through ion conductivity or porosity permitting transport of small ions through its structure. Invariably a divided cell will feature a larger resistance than undivided cells, though this can be minimised by careful design of the membrane and by minimising the inter-electrode distance [53].

Different architectures are possible for divided cells depending on the scale of operations. The most commonly employed is the H-cell (Figure 5.8C), where the two compartments are separated via a narrow flange with an ion-conductive membrane in between. The narrow flange helps to keep costs low by facilitating conductivity between anode and cathode with only a small piece of conductive membrane. The basic H-cell design can be manipulated in order to accommodate the needs of the user. Adding an additional flange to the H-cell can introduce a GDE in place of a foil electrode [354], or fritted glass tubing can be incorporated to allow the user to directly purge the compartment with CO_2 [340].

Choice of membrane: In many cases, the choice of membrane separator is defined simply by the choice of electrolyte; acidic electrolytes operate with a cation exchange membrane (CEM), whereas basic electrolytes operate with an anion exchange membrane (AEM). However, membrane choices in CO_2RR reactors are complicated by the CO_2 transport within the cell. HCO_3^- and CO_3^{2-} formed as part of the aqueous CO_2 equilibrium can cross over AEMs. The carbonates could then be oxidised at the anode to evolve CO_2. This is sometimes referred to as a CO_2 pump, since CO_2 is dissolved in the catholyte and then evolved from the anolyte on the other side of the membrane. CO_2RR products are also susceptible to transport across AEMs. Negatively charged formate ions may cross over in the same way as carbonate ions (Figure 5.9). Neutral liquid-phase products cannot cross over by migration, but they may be drawn across by electroosmotic drag due to the flux of carbonate species towards the anode [355].

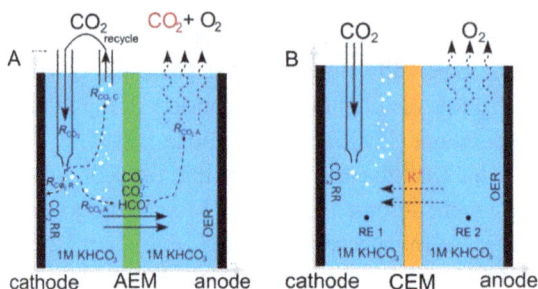

Figure 5.9: Schematic H-cell designs for CO_2RR with an AEM (A) and CEM (B). The AEM facilitates carbonate transfer across the membrane, resulting in CO_2 evolution at the anode as an effective CO_2 pump. The CEM facilitates K^+ transport across the membrane, lowering the conductivity of the anolyte and increasing the cell potential. Adapted with permission from reference [358], Copyright © 2019 American Chemical Society.

A further complication is that the same carbonates can crystallise within the membrane structure, which increases the overall resistivity of the membrane and decreases cell performance [356]. Given that HCO_3^- is also a popular electrolyte anion for the CO_2RR, there is a clear need to hinder the transport of carbonates across the membrane for AEMs to be viable. CEMs are a possible alternative, but these require the use of acidic electrolytes, which necessitate expensive anodic catalysts in order to drive the OER [357]. CEMs are also susceptible to cation transport from the anodic to cathodic compartment, decreasing the conductivity in the anolyte and increasing the overall cell potential [358].

It is worth noting that, even within the same class of membrane, different options can impact the reactor function. Lei et al. demonstrated this by analysing different AEMs, where optimising the choice of membrane gave a ~ 10% increase in formate selectivity [131].

Bipolar membranes: One proposed solution to this has been the development of bipolar ion exchange membranes (BPMs). These membranes are comprised of two phases: one with a cation exchange component and one with an anion exchange component (Figure 5.10). The BPM may be operated in one of two modes. If the anion exchange side faces the cathode, OH^- and H^+ transport from the catholyte and anolyte, respectively, result in water formation within the BPM as a self-hydrating mechanism. Alternatively, if the anion exchange side faces the anode, the opposite reaction occurs, so water electrolysis within the membrane generates OH^- and H^+, which then diffuse out of the BPM into anolyte and catholyte, respectively [359]. The latter orientation is more relevant to CO_2RR, since the active supply of H^+ and OH^- into the electrolyte from water dissociation can help to maintain a constant cell pH, which helps to improve long-term stability [360]. Water dissociation within the BPM can be facilitated with a catalyst such as a metal hydroxide [361], metal oxide [362] or graphene oxide [363], although high-performing BPMs have also been produced without these [364].

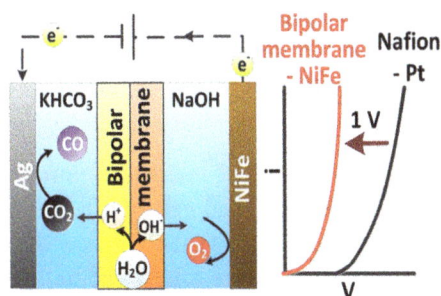

Figure 5.10: Schematic cell design for the CO_2RR featuring a bipolar membrane (BPM). The BPM separates an Ag cathode in neutral $KHCO_3$ from a NiFe anode in basic NaOH. Maintaining conductivity while separately optimising the anolyte and catholyte is able to reduce the cell voltage by ~ 1 V. Reproduced with permission from reference [365], Copyright © 2016, American Chemical Society.

In both orientations, the key part of the BPM is the interface between the CEM and AEM components. Charged species are unable to cross over the entire BPM, since the transport of cations is hindered by the AEM and anions by the CEM component. The net result is hindering unwanted transport across the membrane while maintaining cell conductivity. Importantly, the bipolar electrode allows different anolyte and catholyte to be used, so the catholyte can be optimised to the CO_2RR while the anolyte is separately optimised to the OER.

Of course, BPMs are not without their own drawbacks. Despite their structure, some degree of ion crossover is still possible [366], and they are susceptible to delamination and dehydration, particularly when operating at large current densities [367]. It is also possible for CO_2 to be generated within the BPM at the CEM and AEM interface by reaction of protons from the CEM side with carbonates from the AEM side. Assuming an equal thickness of the CEM and AEM component, this

CO_2 has an equal likelihood of diffusing to the cathodic or anodic side, resulting in some degree of CO_2 pumping across the BPM. This motivated Pătru et al. to create a BPM with an asymmetrical distribution to drive CO_2 towards the cathode instead [368]. Continued supply of protons to the cathode can also promote the parasitic HER. Yan et al. attempted to control this with a BPM containing a weakly acidic CEM component [369]. The weaker acidity of the CEM provided sufficient proton supply for the CO_2RR without accelerating the HER.

5.2.2 Flow cells

Introducing solution flow into a reactor is an effective means to increase mass transport to the electrode surface. This could use flow to increase the supply of CO_2-saturated solution to an electrode for CO_2RR. Alternatively, the motivation could be to provide fresh electrolyte over the course of CO_2RR so that the pH is kept constant, rather than becoming increasing basic as is the case for CO_2RR in static solutions. As with static cells, it is possible to operate flow-by cells in divided and undivided configurations [53]. Broadly speaking, there are two main configurations for flow cells: flow-by and flow-through.

Flow-through cells: Flow-through electrodes can be thought of a subset of porous or mesh electrodes that are employed in a specially designed cell to allow the electrolyte to flow through their porous structure (Figure 5.11). This maximises the interaction of the electrolyte with the electrode structure and prevents the stagnation of electrolysis products within the pores [370]. These designs are more common for areas where total electrolysis in a single pass is very important, such as for electrochemical wastewater treatment [371], although there are CO_2RR applications as

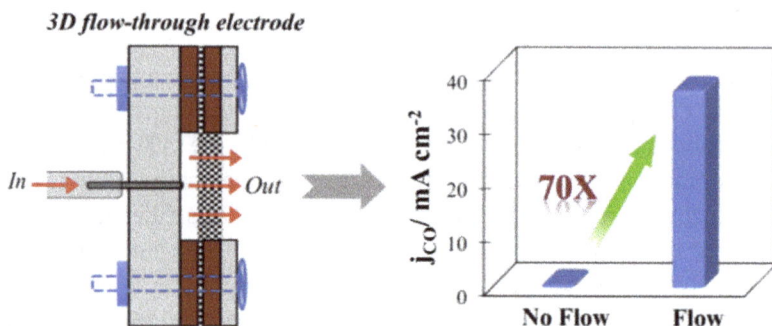

Figure 5.11: Schematic reactor design with a flow-through electrode configuration. CO_2-saturated electrolyte is introduced via the inlet and flowed through a porous cathode. CO_2RR products are then collected in the outflow. Enhanced interaction of CO_2 with the catalyst structure significantly increases the partial current density for CO (j_{CO}). Reproduced with permissions from reference [372], Copyright © 2019 American Chemical Society.

well. Vedharathinam et al. demonstrated this for CO_2 reduction to CO at Ag nanoparticles deposited on an Al foam cathode [372]. Flowing CO_2-saturated ionic liquids through the cathode gave a 70-fold increase in the CO partial current density and 7-fold increase in faradaic efficiency versus static solution.

Flow-by cells: The flow-by cell design involves solution flow perpendicular to the electrode surface (Figure 5.12). Fresh electrolyte is supplied to the catalyst and electrolysis products are carried away, maintaining a relatively constant reaction environment throughout the operation. The simplest design is the one-compartment electrolyser, where CO_2-saturated electrolyte flowed between the parallel plate cathode and anode, positioned on either side of the flow channel. Minimising the resistance of the set-up requires the anode and cathode to be in close proximity, which increases the chance of CO_2RR products from the cathode being re-oxidised to CO_2 at the anode, so it is common to divide the flow cell into a two-compartment electrolyser.

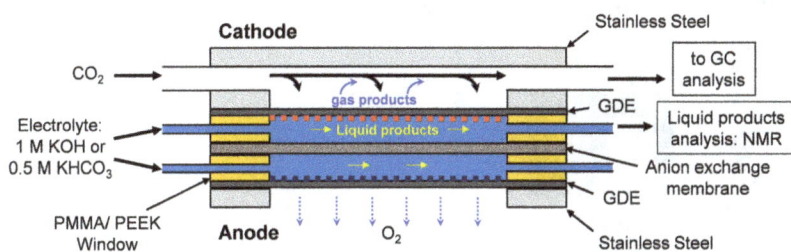

Figure 5.12: Schematic design for a three-compartment flow cell for CO_2RR. The cathode features a gas diffusion electrode with gas flow chamber behind and a catholyte flow chamber in front. The third chamber cycles the anolyte on the other side of an anion exchange membrane. In this case, the anode is also a GDE to permit O_2 escape from the anolyte chamber. Reproduced with permission from reference [376], Copyright © 2015 Elsevier B.V.

Considerations regarding anolyte, catholyte and membrane choice for two-compartment flow cells are the same as for static H-cells. Although solution flow can increase the supply of CO_2-saturated solutions, the overall reaction rate is still somewhat limited by low CO_2 solubility [373]. Most one- and two-compartment CO_2 flow cells employ ionic liquids in order to benefit from the improved CO_2 solubility, as discussed in Section 4.3 [286]. Further enhancements in the CO_2RR rate come from incorporating GDEs into the flow cell set-up [374]. These are termed "three-compartment" cells, with a gas compartment supplying CO_2 to the back of the GDE, a catholyte compartment supplying electrolyte to the GDE surface and an anolyte compartment supplying electrolyte to the anode. Some of the best reported stabilities for CO_2RR have come from three-compartment electrodes, including Dinh et al. producing 70% ethylene at ~ 100 mA cm^{-2} for 150 h [225] and Yang et al. producing 73–91% formic acid at 200 mA cm^{-2} for over 1,000 h [375].

Flow fields: Although basic flow cell schematics usually represent gas and liquid flow down a simple open channel, real flow cells must pay careful attention to the shape and size of the flow channels in order to ensure an optimum distribution of the flow and interaction of flowed species with the electrode surface (Figure 5.13). For example, if CO_2 flows down a very deep flow channel, then some of the gas volume may not interact with the GDE at all due to the prohibitive distance. On the other hand, if the channel is very narrow, then the user will have to deal with higher internal pressures that can increase the chance of leakage or component fracture [53].

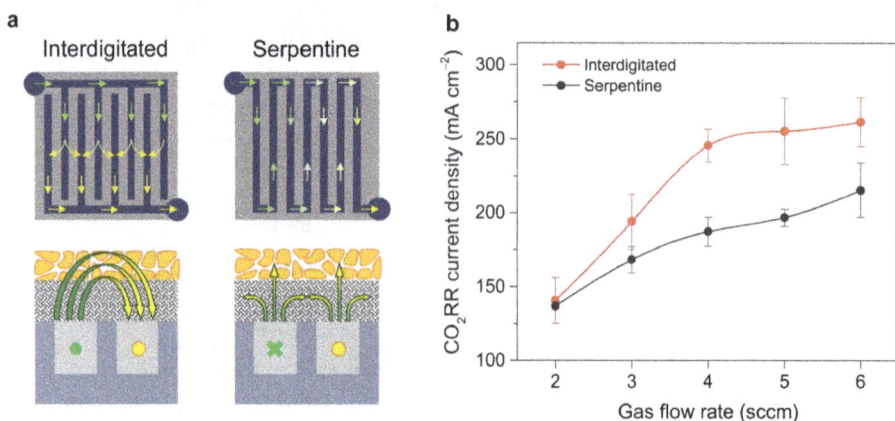

Figure 5.13: (a) Schematic designs for interdigitated versus serpentine flow fields. Top row shows the flow directions in the channels, and bottom row shows the flow within the GDE structure. The interdigitated field can be seen to force more flow through the GDE to give a greater interaction of CO_2 with the catalyst at the GDE surface. (b) Related CO_2RR current densities recorded under the same conditions using an interdigitated (red) or serpentine (black) flow field. The greater current density with an IFF is due to the greater interaction between CO_2 and the catalyst. Reproduced with permission from reference [328], available on Creative Commons CC-BY license, Copyright © 2021 Xing et al.

One of the primary challenges in flow field design is establishing an even flow profile and distribution of pressure across the entirety of the electrode surface. Since CO_2 will begin to be consumed as soon as it enters the flow field, the concentration at the start of the channel will be greater than at the end, creating an asymmetrical current distribution across the electrode [377]. It is often challenging to identify the most efficient flow field design since the flow field geometry is usually varied along with a catalyst material and other parameters, although there is evidence that serpentine designs are some of the best performing across multiple fields [378].

An interesting alternative is the interdigitated flow field (IFF), where a series of parallel channels are interwoven, each with the end sealed off. The sealed ends force flow to move out into the porous GDE structure before returning into the next downstream channel. This essentially acts as a form of flow-through electrode but with a much smaller pressure drop [379, 380]. IFFs are more commonly employed in fuel cell technologies but some CO_2RR groups are beginning to take advantage of the high rates of reaction and high single-pass conversion rates that they provide. One of the best examples is for the related CO reduction to acetate, where Ripatti et al. achieved 68% conversion on a single pass, producing working solutions of >1 M acetate in the product stream [381].

5.2.3 Zero-gap cells

Zero-gap cells are a class of electrolysers that are designed to bring the anode and cathode as close to each other as possible, thereby lowering the cell resistance and giving a more energy-efficient cell. The primary design is the same as proton exchange membrane (PEM) fuel cells, where the anode and cathode are pressed against the same ion exchange membrane, which acts as the only separator between the two electrodes. These are often referred to as membrane electrode assemblies, since the anode, cathode and membrane are combined into a single component for insertion into the reactor [382]. Reactants are then introduced to the electrodes via flow channels behind the respective GDEs. Water for CO_2RR is introduced by humidifying the CO_2 gas before it is flowed behind the GDE. As well as low cell resistance, this design offers the advantages of providing water for CO_2RR without dissolving CO_2 to negate solubility challenges, and removes the need for catholyte pumps to lower the overall cost [383].

Product distribution: It is worth noting that this water supply is somewhat limited, and so zero-gap cells are generally only able to produce two-electron CO_2RR products such as CO and formate in significant quantities [36]. It is possible to control the amount of water in CO_2 by bubbling CO_2 in a thermostatically controlled humidifier, where higher temperatures provide more water vapour in the CO_2 stream. By optimising the vapour supply, Lee et al. were able to record 93% formate at ~ 50 mA cm^{-2} [384]. Fan et al. introduced water in vapour form through a solid porous layer in between the cathode and anode, separated by an AEM and CEM, respectively [385]. CO_2RR at the cathode produced formate, which combined with protons from H_2 oxidation at the anode to give concentrated formate in the vapour stream. De Mot et al. took a different approach by directly injecting water into the CO_2 steam using an HPLC pump [386]. The authors calculated the amount of water consumed by CO_2RR and lost by electroosmotic drag and injected that precise amount, giving 80% formate at 100 mA cm^{-2} while hindering bicarbonate crystallisation to extend the cell lifetime.

Further increases have been possible by tuning the interaction between CO_2 and the membrane, since it is in direct contact with the cathode catalyst [387]. Kutz et al. reported 98% CO at >100 mA cm^{-2} over 6 months of continuous reduction [388]. The activity and stability were attributed to their novel imidazolium-based membrane, which contained functional units based on imidazolium ionic liquids that are designed to enhance CO_2RR. The same group was able to use the same concept to produce 94% formate at 140 mA cm^{-2} over 550 h by moving from Ag to Sn nanoparticles at the cathode [389].

Hybrid zero-gap cells: In order to access high-order CO_2RR products, certain modifications can be applied to allow zero-gap technologies. One commonly employed technique is to introduce a thin "buffer layer" of electrolyte at the cathode side. This provides an electrolyte to facilitate the proton transfer necessary for CO_2RR, while the anodic side remains zero gap to keep the cell resistance low. Further advancements such as minimising the buffer layer thickness to keep the cell resistance low have allowed up to ~ 60% ethylene at current densities >1.5 A cm^{-2} [339]. The same concept has also been used to enhance the production of formate at SnO_2 catalysts, giving up to 90% formate at an impressive 500 mA cm^{-2} (Figure 5.14) [390].

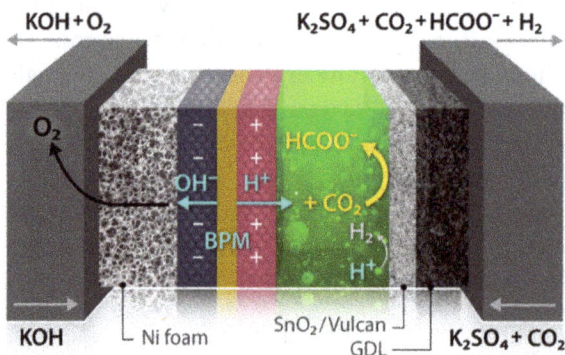

Figure 5.14: Schematic diagram for a hybrid zero-gap CO_2RR cell. On the left, a Ni foam anode is in direct contact with the BPM in a zero-gap configuration. On the right, a 1.27 mm thick catholyte layer (green) is introduced between the BPM and the SnO_2 catalyst layer. This provides the necessary proton transfers for formate product while minimising the cell resistance. Reproduced with permission from reference [390], Copyright © 2020 American Chemical Society.

5.2.4 Microfluidic devices

Microfluidic devices offer a means to efficiently separate the anodic and cathodic reaction environments without the need for membranes or separators in the cell design. Instead, separation is achieved through the characteristic flow profiles of microfluidic channels. Specifically, maintaining a laminar flow profile gives limited mixing between the anodic and cathodic sides of the channels [391].

Separation of anolyte and catholyte: One of the key motivations for separating anolyte and catholyte in CO_2RR cells is to prevent the re-oxidation of liquid-phase CO_2RR products at the anode. In the same way, diverging the flow channel into anodic and cathodic sides downstream from the electrode exploits the laminar flow profile for efficient product separation; CO_2RR products remain close to the cathode side, and so selectively flow into the catholyte outlet in high concentration [392].

It is worth noting that electrolyte separation through laminar flow can also be used to benefit the initial CO_2RR. For instance, using separate inlets into a single laminar flow channel can introduce different anolyte and catholyte into the same microfluidic device (Figure 5.15) [393]. This would allow the user to employ different pH anolyte and catholyte, offering the same kinetic advantages as achieved when using different pH electrolytes with a BPM [394]. Further improvements in turnover rate can be achieved by simply increasing the path length or connecting multiple channels in parallel, providing a relatively straightforward route to up-scaling, although maintaining a uniform flow distribution in larger stacked electrolysers may prove challenging [36, 395].

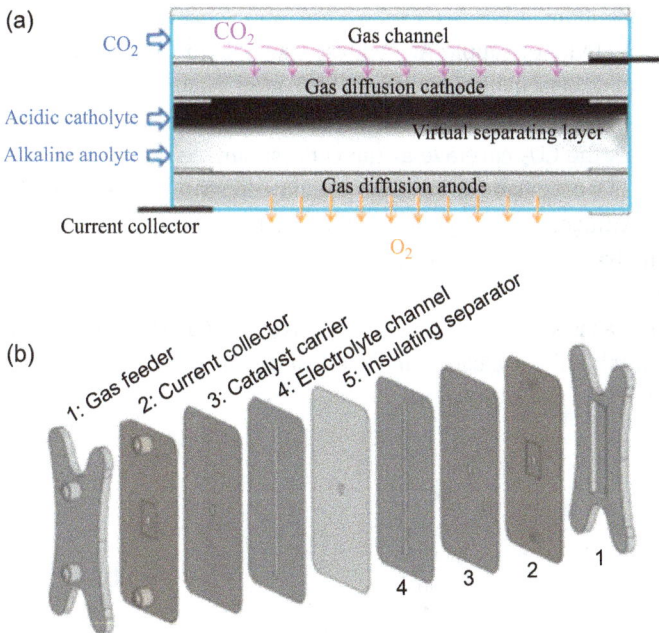

Figure 5.15: Schematic microfluidic reactor for CO_2RR using different anolyte and catholyte. The acidic catholyte and alkaline anolyte are separately chosen to favour CO_2RR and OER, respectively. Reproduced with permission from reference [393], Copyright © 2016 Elsevier Ltd.

Considerations: The primary concern when operating microfluidic devices for any system is maintaining laminar flow. In the case of CO_2RR, it is important to consider the impact of electrolyte flow rate (or catholyte and anolyte flow rate if different electrolytes are employed) and also the CO_2 gas flow rate. The flow profiles will also vary with different microfluidic designs, particularly focusing on channel width and depth [394]. It is also important to consider that some degree of mixing is possible even in laminar flow. Extended operations using different pH anolyte and catholyte could therefore lead to neutralisation at the interface requiring wasteful replacement of electrolytes [396]. Complimentary computational experiments are therefore always recommended when designing a new device in order to ensure that the laminar flow is observed under operating conditions. A further consideration for microfluidic devices for CO_2RR is how to minimise bubble formation. Bubbles are resistive and block adsorption sites, which will slow the rate of CO_2RR [397]. Also, bubble formation during electrolysis can generate turbulence when they detach from the electrode, which will disrupt the laminar flow profile [398].

5.2.5 Reactor pressure, CO_2 partial pressure and temperature

Reactor pressure: Experimentally, it has been shown that an increase in reactor pressure can increase the selectivity of simple Ag catalyst materials to CO with a corresponding decrease in formate selectivity (Figure 5.16) [399, 400]. A number of different causes for this dependency have been proposed, based on the known factor that increased pressure increases the CO_2 coverage at the catalyst surface. Hara et al. proposed that the increased CO_2 coverage gave a corresponding decrease in adsorbed *H, which disfavoured the hydrogenation steps in the formate mechanism to favour CO production [401]. Alternatively, Gabardo et al. proposed that the greatly increased CO_2 concentration at the electrode surface could favour CO production via the reaction between two CO_2 molecules, as is seen in organic aprotic solvents [402]. Of course, these proposed causes are not mutually exclusive, and the real origins of this pressure dependence may be a combination of both factors.

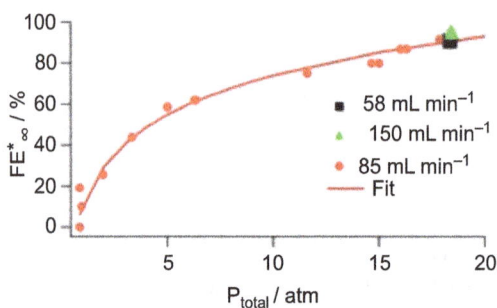

Figure 5.16: Experimental data and computational fit for a pressurised CO_2RR electrolysis cell at 225 mA cm^{-2}. CO_2 flow rates were 58 (black square), 85 (red circle) and 150 (green triangle) mL min^{-1}. Increase in reactor pressure gives a steady increase in faradaic efficiency for CO. Reproduced with permission from reference [399], Copyright © 2014, Springer Science Business Media Dordrecht.

CO₂ partial pressure: The availability of CO_2 at the electrode surface is key in promoting CO_2 reduction to C_{2+} products. Many of the surface modifications discussed in Section 3.3 focus on packing CO_2 close enough to facilitate C–C bonds in order to form ethylene, or hindering CO_2 interactions to form methane. Controlling the local availability of CO_2 has also been achieved by lowering the CO_2 partial pressure in the CO_2 feed gas (Figure 5.17). Wang et al. demonstrated this by diluting the CO_2 input with N_2, where a dilution to 75% CO_2 increased the faradaic efficiency of methane from 8% to 48% at 225 mA cm^{-2} [403]. As well as decreasing the chance for C–C bond formation, the decreased CO_2 coverage favours the protonation of *CO to *CHO, driving CO₂RR down the methane pathway.

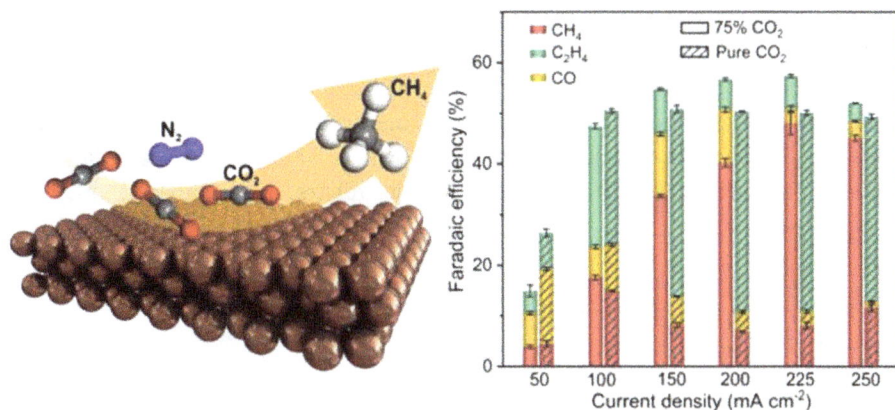

Figure 5.17: (Left) Schematic Cu electrode surface with a CO_2 feed diluted with N_2. (Right) Faradaic efficiency for methane (red), ethylene (green) and CO (yellow) at a Cu electrode when using a pure CO_2 (hashed boxes) and a 75% diluted (open boxes) gas feed. Dilution of the CO_2 feed gives a sizeable increase in methane and corresponding decrease in ethylene. Reproduced with permission from reference [403], Copyright © 2020, American Chemical Society.

Computational studies have shown that an increase in CO_2 partial pressure increases the production of CO at Ag electrodes due to increased CO_2 surface coverage [404]. Interestingly, Moradzaman et al. found that low partial pressures of CO_2 increased the faradaic efficiency for both methane and ethylene if low concentrations of bicarbonate electrolyte are employed [405]. They attributed this trend to the low CO_2 partial pressure increasing the local pH at the catalyst surface, which was only seen if the electrolyte was only weakly buffering. Song et al. found that low partial pressures of CO_2 improved ethylene faradaic efficiencies when operating at lower overpotentials, which they attributed to unreduced CO_2 at the catalyst surface blocking *CO dimerisation [406]. As the overpotential increased, the trend shifted so that concentrated CO_2 streams became more efficient. Clearly, a full consideration of the entire reaction system is essential when changing CO_2 partial pressure to impact CO₂RR product selectivity.

Temperature: Cell temperature can be simply varied in most electrochemical reactors via a thermostatically controlled jacket or water bath. Since CO_2 solubility is temperature dependent, this provides an additional means of varying the CO_2 availability to the catalyst. This is particularly prevalent when using organic solvents such as methanol, since CO_2 is four times more soluble in methanol than water at room temperature, which increases to more than seven times more soluble when the temperature is lower than 273 K [407].

A number of experimental studies have shown that low temperatures increase the selectivity of CO_2RR to methane as the dominant product (Figure 5.18) [408]. Although the mechanism for this mode of action remains unclear, it has been suggested that lower temperatures suppress the thermal excitation of electrons in the catalyst, favouring the formation of the most energetically favourable product [409]. This has a number of direct implications for methane production; CO is more energetically favourable than $HCOO^-$ [74] and adsorbed *CO is more energetically favourable than free CO [94], which favours further reduction steps. Computational studies have also suggested that methane is the most energetically favoured product from CO reduction in near-neutral pH electrolytes, more so than methanol or C_{2+} species [410].

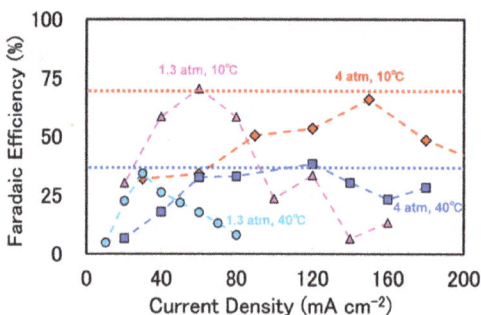

Figure 5.18: Variation in the faradaic efficiency for methane over a range of current densities at combinations of two different temperatures and pressures. The red and blue dotted lines represent the maximum faradaic efficiency achieved at 10 and 40 °C, respectively. Reproduced with permission from reference [408], Copyright © 2016, American Chemical Society.

Choices regarding CO_2 solubility must be considered against diffusion coefficients; low temperatures give high solubility but slow diffusion whereas high temperatures give low solubility but fast diffusion. Löwe et al. investigated formate production at SnO_2 GDEs over a range of temperatures [411]. They found that an intermediate temperature of 50 °C gave the best compromise between solubility and mass transport, reaching 80% formate up to 1 A cm^{-2}.

Unified approach: Of course, the parametric approach to reactor design should consider all of these parameters in tandem, taking into account how they impact the CO_2RR and how they interact with each other [409]. Hashiba et al. demonstrated this for methane production at a Cu electrode [408]. The authors took a combinatorial approach by varying the reactor temperature, pressure and stirring speed (stirring at a regular rpm controlled the rate of mass transport to the electrode). The authors observed that changes to CO_2 pressure and stirring speed increased methane production rates, while lowering the temperature improved selectivity. Combining the benefits of all features produced ~ 70% methane up to ~ 150 mA cm^{-2}.

6 Future perspectives

6.1 High-priority research focus

6.1.1 Extended operations

A sizeable portion of the CO_2RR literature is devoted to material analysis on the 1-h timescale (Figure 6.1) [48]. These studies are useful to material design and refinement, particularly when conducting a parametric analysis of the impact of variations in nanostructure, loading or alloyed element content. However, it is important to acknowledge that materials that perform well on these scales do not necessarily transfer to scaled-up operations. This is particularly true for nanostructured materials that feature high energy active sites, which are susceptible to dissolution and re-deposition during continued operations [197].

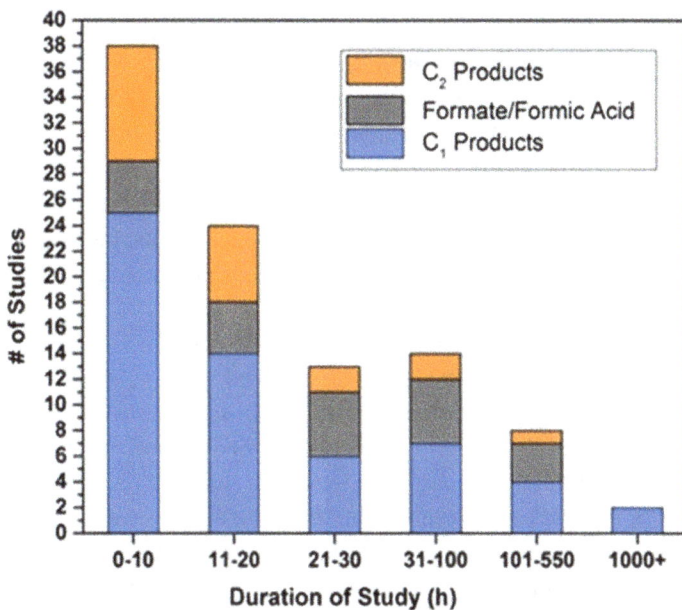

Figure 6.1: Survey of CO_2RR durability studies as reported in the literature between 2013 and 2019. Reproduced with permission from reference [48], Copyright © 2019 Wiley-VCH Verlag GmbH & Co. KGaA, Weinheim.

In order to move CO_2RR out of the lab into wider applications, it is essential to investigate the catalyst, scaffold, membrane and reactor designs over thousands of hours of continued operations. Some reactor designs have already achieved this. One report from Kutz et al. demonstrated stable CO production at 200 mA cm^{-2} over a period

https://doi.org/10.1515/9781501522239-006

of 6 months, with a projected lifetime of 4 years based on increases in cell resistance over that time [388]. However, this exceptionally long-term study is an exception rather than the standard in CO_2RR literature. Studies on the order of 1–7 days are slightly more common and have demonstrated for syngas [42] and ethylene production [225].

It is worth noting that lifetime studies should always be considered alongside operating conditions. Simple H-cell studies can often show high stabilities, thanks to stable foil electrodes and the absence of porous GDEs that can be flooded in flow cells or MEAs. However, less stable MEAs would likely be more economically feasible, thanks to greatly accelerated turnover rates, despite the added costs from component replacement.

An alternative focus of CO_2RR research could look into ways to regenerated degraded reactor components. This is not ideal, as production would most likely need to cease for the regeneration to take place, but this is preferable to the costs and time associated with a need to dismantle a reactor, replace components and reassemble. A key target for component recovery is GDE, which suffers from failure due to flooding, electrowetting and crystallisation of electrolyte within the GDE structure.

Prevention methods have been discussed in Section 5.1.3, but there is still an interest in recovering flooded GDEs, particularly with a view to being able to use cheaper, less hydrophobic scaffolds. The most commonly employed recovery method is to rinse the GDE with deionised water to remove salt crystals, which extended the lifetime, although this could not fully recover the starting performance [412]. It is not ideal to have to dismantle, rinse, dry and reassemble reactor components, but similar improvements have been achieved by flushing the system with deionised water, which is promising for in situ treatment [101, 413].

A further failure mode is catalyst poisoning, which is a key concern for all types of CO_2RR reactor. Poisoning species adsorb onto the catalyst surface, which can block the adsorption of CO_2 and its reduction intermediates. In some cases, catalysts can be recovered by electrochemically stripping the adsorbed poisoning species (Figure 6.2). Ma demonstrated this for an Ag/Ag_2CO_3 catalyst that lost performance after ~ 100 h. By applying an oxidising potential for 3 min, the authors were able to recover performance for a further 60 h [49]. Lee et al. used just a 10 s anodic pulse to remove poisoning CO from a Pd catalyst, enabling constant formate production for over 45 h. The same approach has been used to recover catalytically active sites at Cu and Pd catalysts for ethylene and formate formation, respectively, using a pulsed potential waveform [414, 415]. Regular anodic pulses re-oxidised the Cu surface to recover active surface structures and give a more constant ethylene production for over 8 h.

Alternatively, possible poisoning could be removed before being introduced to the catalyst through careful filtering of gas- and liquid-phase flow media [417]. This negates the need for a recovery step but would of course bring extra costs and complexity to the reactor design.

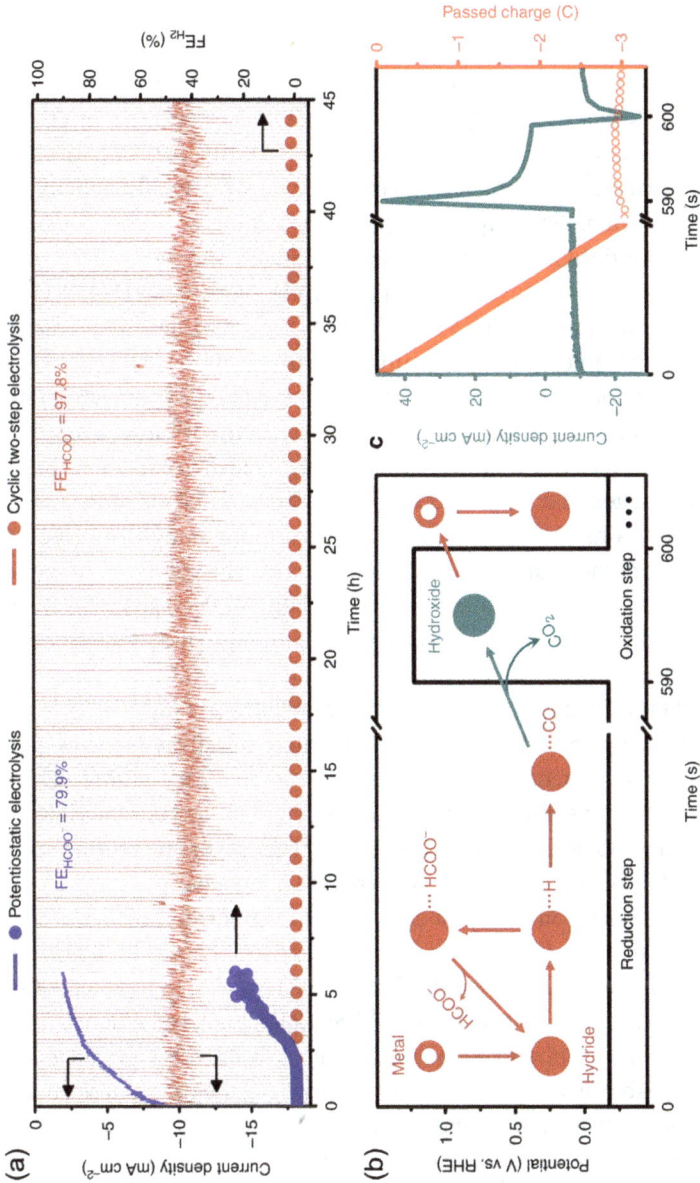

Figure 6.2: (a) Current density and faradaic efficiency for H_2 production using constant potential electrolysis (blue) or with a 10 s anodic pulse for every 10 min (red). The anodic pulse maintains a constant current density and hinders parasitic H_2 production by removing adsorbed CO. (b) Surface reactions were ongoing during the corresponding points in the potential waveform. During reduction, CO_2 is reduced to formate with some parasitic CO. The anodic pulse oxidises the CO to recover the catalytic sites. (c) Current density and charge versus time during the potential pulses. Reproduced with permission from reference [416], available open access on Creative Commons CC-BY license, Copyright © 2019 Lee et al.

6.1.2 Enhanced selectivity

The need for further development of CO_2RR product selectivity is very different depending on the products that are being looked at (Figure 6.3). Synthesis of CO already exceeds 95% in a number of published examples [103, 105, 112, 114], and the requirement for only two electron transfers means that many top catalysts have an inherently high turnover rate. Similarly for ethylene production, the latest developments in GDE and reactor design have enabled even relatively simple Cu electrodes to produce high proportions of ethylene at large current densities [177–179]. Recently published examples producing ~ 45% ethylene at current densities greater than $1 \, A \, cm^{-2}$ represent a significant step towards industrial viability [339].

Other products have not yet reached these impressive standards. Research into methane production focused on how to harness the selectivity of Cu as a starting point, while driving the selectivity towards methane only by disfavouring C–C bond formation. Isolating Cu sites in a disperse catalyst has been an effective means of achieving this [164], but a dispersed catalyst will have an inherently low turnover rate due to the limiting number of catalytically active sites. The future of methane electrocatalysts will likely focus on tandem catalysts, where catalytic space between isolated Cu generates *CO from the two electron reduction of CO_2 so that catalyst space and loading are not wasted [208].

Alternatively, isolating Cu crystal facets that favour methane has provided promising selectivity. The challenge there will be maintaining these selectivity facets by preventing surface dissolution or reorganisations that are commonplace during CO_2RR [160]. Single-crystal applications also do not offer protection from the parasitic HER, so these catalysts would have to be combined with either a chemical or engineering solution to lower the rate of water reduction to reach higher reaction rates.

Regarding liquid products, only formate has shown a significant number of systems capable of synthesis from CO_2RR with both high selectivity and high current density. Optimised Sn-based catalysts combined with GDEs are able to produce >90% formate at current densities around $300 \, mA \, cm^{-2}$ [154]. Methanol and ethanol have not yet been produced in such sizeable amounts. In most cases, these products are produced alongside sizeable amount of methane and ethylene, their gas-phase equivalents. Shifting the selectivity towards liquid products is highly surface dependent, as demonstrated by a number of authors who have shifted the selectivity with only minor change to the doping levels in copper alloys [78, 191].

Although there is still some uncertainty in the exact mechanistic cause of the deviation between the methane/methanol and ethylene/ethanol routes, computational studies have indicated that the presence of surface-bound *H is important in driving the hydrogenation of CO_2RR intermediates to give the alcohol. The key here would be to provide sufficient quantities of *H to give the alcohol product without reducing the density of the methanol/ethanol selective catalyst, particularly for ethanol which needs neighbouring sites of C–C bond formation [78].

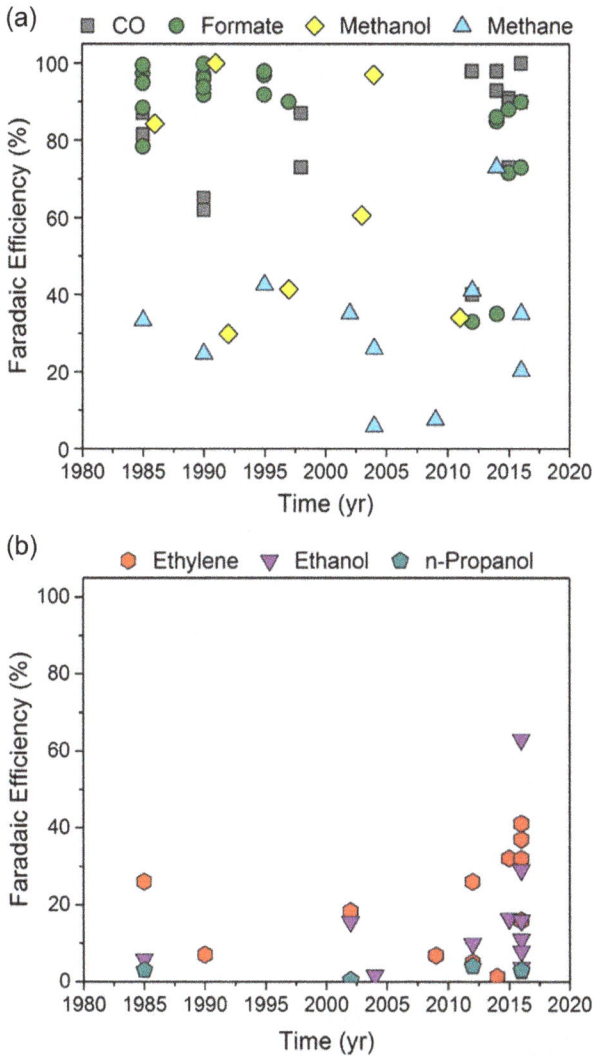

Figure 6.3: Developments in CO_2 faradaic efficiency for a number of CO_2RR products between 1985 and 2018. Lower order CO_2RR products such as CO and formate have consistently reached higher current densities than higher order C_{2+} species. Reproduced with permission from reference [44], Copyright © 2018, American Chemical Society.

Presently, leading materials for C_{3+} CO_2RR products are the furthest behind, with very few examples reaching 20% even when discussing all C_{3+} products combined. The challenge is fundamentally based on how to continually form C–C bonds without intermediate products such as ethylene or ethanol being produced. Increasing the binding energy of the intermediates is unlikely to be feasible as further increases increase the risk of the poisoning of catalytic sites by CO. Instead, some of the leading materials have focused on physically trapping CO_2 and intermediates close to the electrode surface [196]. This permits the re-adsorption of CO_2RR intermediates for further reduction, particularly CO, and increases the local pH during CO_2RR to help hinder the HER. Methods of achieving this through nanoporous materials could be further enhanced with surface adsorbates of polymer coatings if C_{3+} production is to reach the levels of the other leading CO_2RR products.

6.1.3 Considering the anode

A cursory search of the CO_2RR literature reveals that most works give little attention to the choice of anode material. Most works can be described as "half-cell" studies, where investigative techniques are used to optimise the CO_2RR cathode, and the anode is assumed to balance the charge efficiently via OER. Commonly employed materials are divided based on the scale of the investigated cathode. Small-scale liquid cells and GDE half cells tend to use a Pt gauze, which offer a high surface area and kinetically facile OER [341]. Larger GDEs and/or MEAs often employ IrO_2, which offers excellent OER kinetics and facile dispersion onto a GDE or anode plate [101]. However, both Pt and IrO_2 present challenges regarding cost and material scarcity [418].

Despite being so widely used, the OER is a famously slow reaction hindered by sluggish electrochemical kinetics [419]. An alternative to this is to use the hydrogen oxidation reaction (HOR) at the anode, which is considerably faster [385]. This can be advantageous as the more kinetically facile HOR gives a much lower cell potential versus the same cell with OER at the anode [420]. However, the gains in terms of cell potential must be weighed against the additional costs associated with H_2 sequestration, transport and storage, as well as the well-known hazards associated with compressed H_2. Future developments in the electrochemical production of H_2 from water electrolysis may help to mitigate some of these costs; however, the present state of technology does not make this route viable [421].

An alternative to the OER or HOR is to use the anode for a second desirable electrochemical reaction so that the reactor produces two useful products for the same charge passed. This is termed a paired electrochemical process and is receiving increasing interest in the electrochemical literature [60]. A well-known example of such a process is the chlor-alkali process, where Cl_2 and NaOH are simultaneously produced via the electrolysis of brine [422]. Many popular options involve the oxidation of organics at the anode coupled to a desirable cathodic process. Recently published examples include biomass oxidation to value-added materials [423], alkane dehydrogenation [424] and the production of useful oxidants such as H_2O_2 [425, 426], HClO and $S_2O_8^-$ [427].

Although relatively unexplored, some successes have already been in this area pairing value-added oxidation with CO_2RR to CO. Llorente et al. demonstrated this for lignin oxidation via a Ce mediator [429], and Pérez-Gallent et al. paired CO_2RR

Table 6.1: Potential oxidation reactions that could be paired with CO_2 reduction in a paired electrochemical process reactor, based on an initial study conducted in reference [60].

Reactant	Price ($ kg^{-1})	Product	Price ($ kg^{-1})	E^0 vs SHE (V)	Production (Mt year^{-1})
Deionised water	0.02–0.1	Oxygen	0.024–0.04	1.23	2
		Hydrogen peroxide	0.56–0.58	1.78	2.8
Ethanol	0.4	Acetaldehyde	1	0.193	1.7
		Acetic acid	0.68–0.92	−0.334	10
		Ethyl acetate	1.21–1.8	−0.208	2.7
1,3-Propanediol	2.2	Acrylic acid	2.25–2.88	0.248	5.2
1,2-Propanediol	1.50–2.05	Lactic acid	1.58–1.87	−0.334	0.45
Glycerol	0.16–0.80			0.041	
		Glyceraldehyde	2.11	0.35	
		Dihydroxyacetone	2.0	0.33	0.004
Benzyl alcohol	1.92–3.47	Benzaldehyde	1.18–2.11	0.193	0.09
		Benzoic acid	1.85	−0.334	0.13
Hydroxymethylfurfural	1.03	2,5-Furan-dicarboxylic acid	32–580	−0.780	0.5
Isopropanol	1.26–1.6	Acetone	0.9–1.28	0.054	6.4
Methanol	0.3	Formaldehyde	0.37–0.74	0.465	18
Methanol	0.34–0.49	Formic acid	0.97–1.08	−0.258	0.95
Ethylene glycol	0.83–1	Glycolic acid	1.84	−0.334	0.04
		Oxalic acid	1.4	−0.455	0.19
Brine	0.05	Chlorine	0.25	1.36	63

with 1,2-propanediol oxidation to lactic acid [430]. Importantly, this latter work reported a 35% decrease in energy consumption versus separate processes, which highlights the motivations of paired electrochemical processes in reducing operating costs. Of course, given the broad scope of plausible reactions listed in Table 6.1, all efforts into paired electrochemical processes must show as much emphasis onto anode design as cathode. Additional considerations to anolyte/catholyte separation may also be necessary in cases where anodic feedstock, products or intermediates could interfere with CO_2RR. This could be achieved through bipolar membrane assemblies or microfluidics depending on the needs of the individual reaction.

6.2 Complimentary technologies

6.2.1 Green energy security

Green energy sources such as solar and wind struggle with intermittent power outputs. During peak operating hours (daylight for solar, windy weather for turbines), they produce excess energy, whereas during off-peak hours they have a deficit. Any widespread application of intermittent green energy sources therefore needs a storage solution to contain the excess energy produced during peak hours, which can then be fed back into the grid to supplement the energy supply during off-peak hours. Decentralised CO_2 reduction is a promising candidate for this application (Figure 6.4). Excess energy during peak operation could be used to power an electrochemical reactor that converts CO_2 into a biofuel. This fuel would then be consumed during off-peak hours to supplement the energy grid [43].

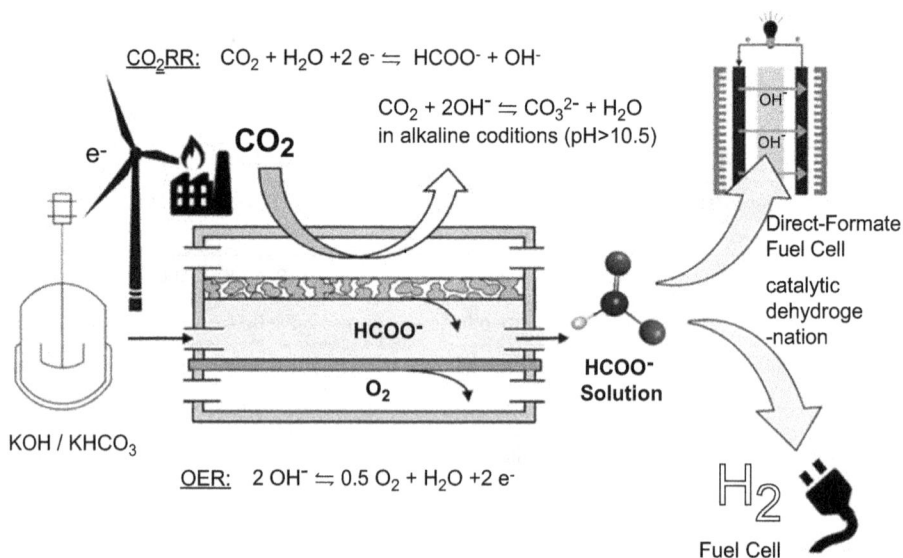

Figure 6.4: Schematic diagram for a CO_2RR reactor to provide green energy security. Excess power from a wind turbine converts CO_2 into formate, which can then either be fed into a direct formate fuel cell or dehydrogenated into H_2 as a secondary fuel source. Reproduced with permission from reference [432], Copyright © 2019 WILEY-VCH Verlag GmbH & Co. KGaA, Weinheim.

The first decision here is that of the many CO_2RR products which should be the biofuel target. Liquid products seem to be more desirable than gas-phase products for this application, as the product could be collected and stored in an electrolyte reservoir, rather than needing capture and compression in gas storage tanks. Formate, methanol or ethanol therefore appear to be the most viable species to fill this role. A further benefit to these species is the existence of complementary fuel cell technologies that are designed to convert these biofuels into energy. The goal is to produce a closed-loop technology featuring the electricity–biofuel–electricity cycle [431].

Of these, formate would likely be realised the soonest, thanks to the higher faradaic efficiencies currently accessible for formate production from CO_2RR technologies with respect to those of methanol and ethanol. The key requirement is to produce a suitable concentrated formate solution during peak operation so that the formate fuel cell can efficiently produce electricity as needed. Typically, direct formate fuel cells operate with >1 M formate as a benchmark value [433]. This concept was demonstrated by Xiang et al. who were able to produce 0.5 M formate during a 1-h electrolysis, which could then be fed into a direct formate fuel cell for energy generation [434]. Recent work from Xia et al. produced a solid-electrolyte device capable of producing >1.5 M formate in the liquid phase, or >12 M in the vapour phase, indicating that research in this area is rapidly approaching viability [435].

Alternatively, formate may be dehydrogenated to give H_2, which could be used as a fuel itself in a PEM fuel cell [432]. The choice here involves a compromise between the simplicity of the closed-loop design versus the technological readiness level of the fuel cell technology. H_2-based fuel cells are presently more advanced but the need to dehydrogenate formate and capture and supply H_2 into a secondary reactor adds cost and complexity to the system [436]. Continuing developments in the direct formate fuel cell coupled with CO_2RR reactors capable of increasing high concentrations of formate are of great interest as researchers look to make closed-loop green energy security economically viable.

6.2.2 Multi-synthesis reactors

In many cases, the reduction of CO_2 to any one of its products will not be the end of the synthesis chain. Many of the CO_2RR products are attracting high research interest due to what they can be converted into, rather than a direct use of the species itself. Key examples include ethylene, which can be oxidised to ethylene oxide for poly(ethylene glycol) synthesis [437], and syngas, which can be converted into a broad range of liquid hydrocarbons via Fisher–Tropsch synthesis [18].

When considering the complete financial, energy and carbon costs of synthesis processes involving CO_2RR, transport of reaction intermediates from a CO_2RR reactor to a secondary plant for conversion will always be a significant contributor, particular to the carbon costs when transporting large volumes of liquid chemicals. There is therefore an interest in hyphenating CO_2RR reactors with a secondary reactor capable of an online conversion of the CO_2RR product into the final target (Figure 6.5). This is a similar concept to tandem catalysis, where one catalyst site reduces CO_2 to CO, with a second site tuned to CO reduction to higher order CO_2RR products (Figure 6.6) [202].

Moving the secondary reaction step to a separate reactor gives greater reaction control, and therefore gives access to a much broader range of products.

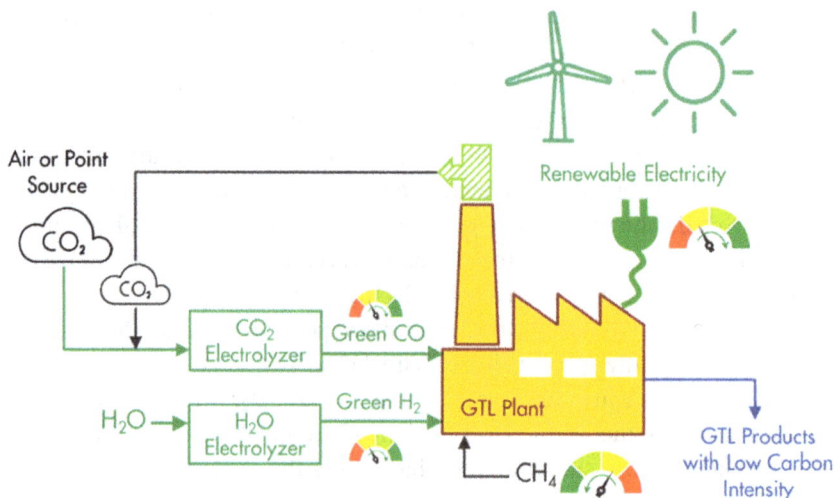

Figure 6.5: Schematic design for a gas-to-liquid/power-to-liquid process, where H_2O and CO_2 electrolysers provide CO and H_2 into a secondary reactor that is capable of producing value-added liquid products via a conventional Fisher–Tropsch synthesis. Methane is also depicted as an alternative source of syngas. Developments in CO_2RR electrolysers could allow the H_2O electrolyser and methane supply to be omitted, with a targeted $CO:H_2$ ratio coming directly from one single electrolyser. Reproduced with permission from reference [438], Copyright © 2020 American Chemical Society.

Of the potential targets, syngas applications are some of the most promising, thanks to the ability for CO_2RR reactors to tune the $CO:H_2$ ratio in situ for direct feed into the secondary reactor without needing intervention steps such as purification, separation or concentration. Haas et al. demonstrated this with a CO_2RR reactor that is fed into a secondary bioreactor, which converted syngas into either ethanol and acetate or butanol and hexanol depending on the bacteria that were employed [42]. A comparative analysis from Zhou et al. demonstrated that this two-step production route is energetically more efficient than the equivalent one-step process, thanks to the ability to tune the reaction environment to the individual steps [439]. This provides a feasible route of CO_2 to higher order alcohol products that are challenging to produce by purely electrochemical means [440].

Jensen et al. demonstrated that CO from CO_2RR can be fed into a secondary reactor for a broad range of chemical synthesis using Pd-catalysed carbonylations, aminocarbonylation and the Suzuki coupling reactions [441]. The authors demonstrated further hybridisation with a carbon capture device, where CO_2 was captured from air in chamber 1, which was then fed into chamber 2 for CO_2RR and chamber 3

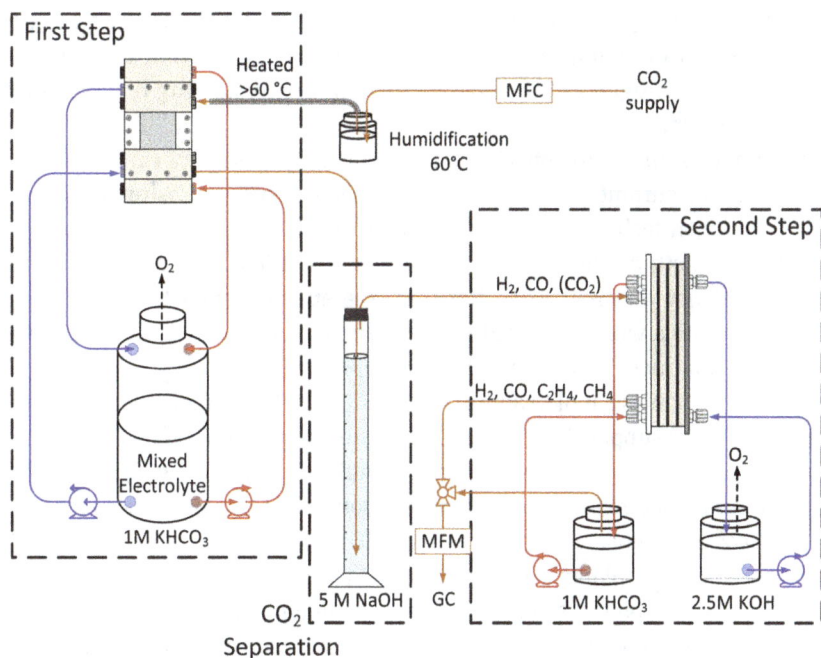

Figure 6.6: Schematic design for a two-step CO_2RR reactor. The first step reduces CO_2 to CO over an Ag catalyst. The CO in the outflow is then humidified in NaOH and fed into a secondary reactor with a Cu catalyst for conversion to higher order CO_2RR products. Reproduced with permission from reference [202], Copyright © 2019 Elsevier Ltd.

for the synthesis step. This is a promising representation of how a delocalised CO_2-based synthesis reactor could function on a larger scale.

Further energy gains could come from combining the benefits of multi-synthesis reactors with paired electrochemical processes. Reactor 1 would convert CO_2 to one intermediate species at the cathode, while the anode carries out an oxidation to produce a second intermediate. These two intermediates then react together to produce the final product. This has been demonstrated by Zhong et al. for the formation of 2-bromoethanol via the simultaneous reduction of CO_2 to ethanol and oxidation of Br^- to B_2 in a membraneless cell [442]. Not all reactions are feasible in a one-pot synthesis such as this one, but the concept is transferrable to a system where the secondary reaction occurs in a customised secondary reactor.

6.2.3 Broader considerations

It is important to note that the challenge to lowering CO_2 levels in order to protect the Earth from global warming will not be solved by one single technological solution. While it is tempting to focus on a "magic bullet" technology, and some technologies indeed offer impressive impacts, a global integration of multiple carbon capture technologies is the only practical means to meeting the 2050 target of net-zero CO_2 emissions.

A key example of this is to consider the movement and usage of carbon, rather than just CO_2, for the current energy, manufacture and transport infrastructures. Employing CO_2 reduction technologies in the manufacture of value-added chemicals is attractive as it can both consume CO_2 before it is released into the atmosphere and lower our dependency on fossil fuels. However, the environmental benefits of any CO_2 reduction synthesis will be negated if the resultant chemical product is shipped to its point of use via trucks and tankers that consume fossil fuels.

Using CO_2 reduction as a tool to address global CO_2 levels will require careful consideration of the entire supply chain from CO_2 capture through end usage. Certain relatively straightforward alterations would have significant impacts, such as building CO_2 reduction infrastructure on-site alongside the source of CO_2 emissions to remove transportation costs and emissions. The choice of CO_2 source should also be carefully considered. Fermentation plants are an attractive first choice, since the CO_2 is relatively clean, so purification costs can be kept to a minimum. Clearly, it is not desirable to undertake significant investment of CO_2 reduction technologies into an industry that could itself be made carbon-neutral in the near future, so a long-term vision is essential.

Hyphenating green energy sources such as solar panels into CO_2 reduction reactors cuts our carbon costs for the energy requirements. Excess energy could be transferred into the grid at peak operating times, or electrochemically generated biofuels could supplement the energy grid under off-peak conditions. This is an area where bifunctional reactors could excel; anodic electrochemical synthesis would produce value-added products for commercial applications, while the cathodic process generates CO_2-sourced biofuels to supplement the intermittent green energy source.

Hyphenating with green energy is an effective way to keep financial and carbon costs down. Further reductions could come from positioning the reactor on site with a viable CO_2 source, such as a biogas or a fermentation plant. Whether via the production of biofuel for energy security or as a viable route to value-added CO_2RR products, decentralised CO_2RR reactors seem likely to reach technological readiness before large-scale industrial applications, thanks to localised production removing costs associated with transport of feedstock gases.

In any case, the implementation of electrochemical CO_2 reduction has the potential for a far-reaching impact. However, reaching this lofty potential requires the entire system to be carefully considered, from surface catalysis, to reactor design and to industrial up-scaling and infrastructure.

References

[1] Álvarez, A., et al. Challenges in the greener production of formates/formic acid, methanol, and DME by heterogeneously catalyzed CO_2 hydrogenation processes, Chem. Rev. 2017, 117, 9804–9838.

[2] Allen, S.K., et al. Summary for policymakers, in: Field, C.B., Dahe, Q., Stocker, T.F., Barros, V. (Eds.) Managing the risks of extreme events and disasters to advance climate change adaptation: Special report of the intergovernmental panel on climate change, Cambridge University Press, Cambridge, 2012, 3–22.

[3] Climate Change Committee, Reducing UK emissions – 2019 progress report to parliament, Climate Change Committee, 2019.

[4] Jones, N. Troubling milestone for CO_2, Nat. Geosci. 2013, 6, 589–589.

[5] Yang, Z., et al. Electrochemical energy storage for green grid, Chem. Rev. 2011, 111, 3577–3613.

[6] Nitopi, S., et al. Progress and perspectives of electrochemical CO_2 reduction on copper in aqueous electrolyte, Chem. Rev. 2019, 119, 7610–7672.

[7] Houghton, R.A. Balancing the global carbon budget, Annu. Rev. Earth Planet Sci. 2007, 35, 313–347.

[8] Le Quéré, C., et al. Global carbon budget 2015, Earth Syst. Sci. Data 2015, 7, 349–396.

[9] Intergovernmental Panel on Climate Change, Climate change 2014: Mitigation of climate change: Working group III contribution to the IPCC fifth assessment report, Cambridge University Press, Cambridge, 2015.

[10] Beck, L. Carbon capture and storage in the USA: The role of us innovation leadership in climate-technology commercialization, Clean Energy 2020, 4, 2–11.

[11] Pletcher, D. The cathodic reduction of carbon dioxide – what can it realistically achieve? A mini review, Electrochem. Commun. 2015, 61, 97–101.

[12] Zhang, X., Guo, S.-X., Gandionco, K.A., Bond, A.M., Zhang, J. Electrocatalytic carbon dioxide reduction: From fundamental principles to catalyst design, Mater. Today Adv. 2020, 7, 100074.

[13] Song, J.T., Song, H., Kim, B., Oh, J. Towards higher rate electrochemical CO_2 conversion: From liquid-phase to gas-phase systems, Catalysts. 2019, 9, 224.

[14] Martino, G., Courty, P., Marcilly, C., Kochloefl, K., Lunsford, J.H. Energy-related catalysis, in: Ertl, G., Knözinger, H., Weitkamp, J. (Eds.) Handbook of heterogeneous catalysis, Wiley-VCH, Weinheim, Germany, 1997, 1801–1900.

[15] Hietala, J., et al. Formic acid, in: Ley, C. (Ed.) Ullmann's encyclopedia of industrial chemistry, Wiley-VCH, Weinheim, Germany, 2000, 1–22.

[16] Bierhals, J. Carbon monoxide, in: Ley, C. (Ed.) Ullmann's encyclopedia of industrial chemistry, Wiley-VCH, Weinheim, Germany, 2000, 1–22.

[17] Küngas, R. Review – electrochemical CO_2 reduction for CO production: Comparison of low- and high-temperature electrolysis technologies, J. Electrochem. Soc. 2020, 167, 044508.

[18] Hernández, S., et al. Syngas production from electrochemical reduction of CO_2: Current status and prospective implementation, Green Chem. 2017, 19, 2326–2346.

[19] Ozer, M., Basha, O.M., Stiegel, G., Morsi, B. Effect of coal nature on the gasification process, in: Wang, T., Stiegel, G. (Eds.) Integrated gasification combined cycle (IGCC) technologies, Woodhead Publishing, Cambridge, UK, 2017, 257–304.

[20] Bulushev, D.A., Ross, J.R.H. Towards sustainable production of formic acid, ChemSusChem 2018, 11, 821–836.

[21] Lee, H.J., et al. Production of H_2-free CO by decomposition of formic acid over ZrO_2 catalysts, Appl. Catal. A 2017, 531, 13–20.

https://doi.org/10.1515/9781501522239-007

[22] Supronowicz, W., et al. Formic acid: A future bridge between the power and chemical industries, Green Chem. 2015, 17, 2904–2911.

[23] Weekly, C. Global market for emission control catalysts to be worth over $16 bn by 2021, Focus Catal. 2016, 2016, 2.

[24] Biswas, S., Kulkarni, A.P., Giddey, S., Bhattacharya, S. A review on synthesis of methane as a pathway for renewable energy storage with a focus on solid oxide electrolytic cell-based processes, Front. Energy Res. 2020, 8, 229.

[25] Din, I.U., Shaharun, M.S., Alotaibi, M.A., Alharthi, A.I., Naeem, A. Recent developments on heterogeneous catalytic CO_2 reduction to methanol, J. CO_2 Util. 2019, 34, 20–33.

[26] Garc'ia, B.L., Weidner, J.W. Review of direct methanol fuel cells, in: White, R.E., Vayenas, C. G., Gamboa-Aldeco, M.E. (Eds.) Modern aspects of electrochemistry no. 40, Springer New York, New York, 2007, 229–284.

[27] Guil-López, R., et al. Methanol synthesis from CO_2: A review of the latest developments in heterogeneous catalysis, Materials. 2019, 12, 3902.

[28] Mohsenzadeh, A., Zamani, A., Taherzadeh, M.J. Bioethylene production from ethanol: A review and techno-economical evaluation, ChemBioEng Rev. 2017, 4, 75–91.

[29] Fan, Q., et al. Electrochemical CO_2 reduction to C_{2+} species: Heterogeneous electrocatalysts, reaction pathways, and optimization strategies, Mater. Today Energy. 2018, 10, 280–301.

[30] Choi, C., et al. Highly active and stable stepped Cu surface for enhanced electrochemical CO_2 reduction to C_2H_4, Nat. Catal. 2020, 3, 804–812.

[31] Bušić, A., et al. Bioethanol production from renewable raw materials and its separation and purification: A review, Food Technol. Biotechnol. 2018, 56, 289–311.

[32] Dias De Oliveira, M.E., Vaughan, B.E., Rykiel, E.J. Ethanol as fuel: Energy, carbon dioxide balances, and ecological footprint, BioScience. 2005, 55, 593–602.

[33] Pal, P., Nayak, J. Acetic acid production and purification: Critical review towards process intensification, Sep. Purif. Rev. 2017, 46, 44–61.

[34] Sawada, H., Murakami, T. Oxalic acid, in: Ley, C. (Ed.) Kirk-Othmer encyclopedia of chemical technology, John Wiley & Sons, Hoboken, NJ, USA, 2000, 882–902.

[35] Riemenschneider, W., Tanifuji, M. Oxalic acid, in: Ley, C. (Ed.) Ullmann's encyclopedia of industrial chemistry, Wiley-VCH, Weinheim, Germany, 2000, 529–543.

[36] Perry, S.C., Leung, P.-K., Wang, L., Ponce de León, C. Developments on carbon dioxide reduction: Their promise, achievements, and challenges, Curr. Opin. Electrochem. 2020, 20, 88–98.

[37] Ooka, H., Figueiredo, M.C., Koper, M.T.M. Competition between hydrogen evolution and carbon dioxide reduction on copper electrodes in mildly acidic media, Langmuir 2017, 33, 9307–9313.

[38] Ma, M., et al. Insights into the carbon balance for CO_2 electroreduction on Cu using gas diffusion electrode reactor designs, Energy Environ. Sci 2020, 13, 977–985.

[39] Schulz, K.G., Riebesell, U., Rost, B., Thoms, S., Zeebe, R.E. Determination of the rate constants for the carbon dioxide to bicarbonate inter-conversion in pH-buffered seawater systems, Mar. Chem 2006, 100, 53–65.

[40] Zhong, H., Fujii, K., Nakano, Y., Jin, F. Effect of CO_2 bubbling into aqueous solutions used for electrochemical reduction of CO_2 for energy conversion and storage, J. Phys. Chem. C 2015, 119, 55–61.

[41] Hori, Y. Electrochemical CO_2 reduction on metal electrodes, in: Vayenas, C.G., White, R.E., Gamboa-Aldeco, M.E. (Eds.) Modern aspects of electrochemistry, Springer New York, New York, 2008, 89–189.

[42] Haas, T., Krause, R., Weber, R., Demler, M., Schmid, G. Technical photosynthesis involving CO_2 electrolysis and fermentation, Nat. Catal 2018, 1, 32–39.

[43] Woo, H.M. Solar-to-chemical and solar-to-fuel production from CO_2 by metabolically engineered microorganisms, Curr. Opin. Biotechnol. 2017, 45, 1–7.

[44] Jouny, M., Luc, W., Jiao, F. General techno-economic analysis of CO_2 electrolysis systems, Ind. Eng. Chem. Res. 2018, 57, 2165–2177.

[45] Ho, M.T., Allinson, G.W., Wiley, D.E. Factors affecting the cost of capture for Australian lignite coal fired power plants, Energy Procedia 2009, 1, 763–770.

[46] Raksajati, A., Ho, M.T., Wiley, D.E. Reducing the cost of CO_2 capture from flue gases using aqueous chemical absorption, Ind. Eng. Chem. Res. 2013, 52, 16887–16901.

[47] Zhang, Q., et al. Carbon capture and utilization of fermentation CO_2: Integrated ethanol fermentation and succinic acid production as an efficient platform, Appl. Energy 2017, 206, 364–371.

[48] Nwabara, U.O., Cofell, E.R., Verma, S., Negro, E., Kenis, P.J.A. Durable cathodes and electrolyzers for the efficient aqueous electrochemical reduction of CO_2, ChemSusChem 2020, 13, 855–875.

[49] Ma, M., Liu, K., Shen, J., Kas, R., Smith, W.A. In situ fabrication and reactivation of highly selective and stable Ag catalysts for electrochemical CO_2 conversion, ACS Energy Lett 2018, 3, 1301–1306.

[50] Guilminot, E., et al. Membrane and active layer degradation upon PEMFC steady-state operation, J. Electrochem. Soc. 2007, 154, B1106.

[51] Wu, J., et al. A review of PEM fuel cell durability: Degradation mechanisms and mitigation strategies, J. Power Sources 2008, 184, 104–119.

[52] Wu, J., Sun, S.-G., Zhou, X.-D. Origin of the performance degradation and implementation of stable tin electrodes for the conversion of CO_2 to fuels, Nano Energy 2016, 27, 225–229.

[53] Perry, S.C., Ponce de León, C., Walsh, F.C. Review – the design, performance and continuing development of electrochemical reactors for clean electrosynthesis, J. Electrochem. Soc. 2020, 167, 155525.

[54] Burchardt, T. An evaluation of electrocatalytic activity and stability for air electrodes, J. Power Sources 2004, 135, 192–197.

[55] Hernandez-Aldave, S., Andreoli, E. Fundamentals of gas diffusion electrodes and electrolysers for carbon dioxide utilisation: Challenges and opportunities, Catalysts 2020, 10, 713.

[56] Pérez-Fortes, M., Tzimas, E. Synthesis of methanol and formic acid, Techno-economic and environmental evaluation of CO_2 utilisation for fuel production, Institute for Energy and Transport (Joint Research Centre), 2016, 1–90.

[57] Rumayor, M., Dominguez-Ramos, A., Perez, P., Irabien, A. A techno-economic evaluation approach to the electrochemical reduction of CO_2 for formic acid manufacture, J. CO2 Util. 2019, 34, 490–499.

[58] Orella, M.J., Brown, S.M., Leonard, M.E., Román-Leshkov, Y., Brushett, F.R. A general technoeconomic model for evaluating emerging electrolytic processes, Energy Technol. 2020, 8, 1900994.

[59] Spurgeon, J.M., Kumar, B. A comparative technoeconomic analysis of pathways for commercial electrochemical CO_2 reduction to liquid products, Energy Environ. Sci. 2018, 11, 1536–1551.

[60] Na, J., et al. General technoeconomic analysis for electrochemical coproduction coupling carbon dioxide reduction with organic oxidation, Nat. Commun. 2019, 10, 5193.

[61] Verma, S., Kim, B., Jhong, H.-R.M., Ma, S., Kenis, P.J.A. A gross-margin model for defining technoeconomic benchmarks in the electroreduction of CO_2, ChemSusChem 2016, 9, 1972–1979.

[62] Pérez-Fortes, M., Schöneberger, J.C., Boulamanti, A., Harrison, G., Tzimas, E. Formic acid synthesis using CO_2 as raw material: Techno-economic and environmental evaluation and market potential, Int. J. Hydrogen Energy 2016, 41, 16444–16462.

[63] Grills, D.C., Lymar, S.V. Radiolytic formation of the carbon dioxide radical anion in acetonitrile revealed by transient IR spectroscopy, Phys. Chem. Chem. Phys. 2018, 20, 10011–10017.

[64] Kortlever, R., Shen, J., Schouten, K.J.P., Calle-Vallejo, F., Koper, M.T.M. Catalysts and reaction pathways for the electrochemical reduction of carbon dioxide, J. Phys. Chem. Lett. 2015, 6, 4073–4082.

[65] Zhang, X., Lei, T., Liu, Y., Qiao, J. Enhancing CO_2 electrolysis to formate on facilely synthesized Bi catalysts at low overpotential, Appl. Catal. B. 2017, 218, 46–50.

[66] Oh, W., Rhee, C.K., Han, J.W., Shong, B. Atomic and molecular adsorption on the Bi(111) surface: Insights into catalytic CO_2 reduction, J. Phys. Chem. C. 2018, 122, 23084–23090.

[67] Pander, J.E., Baruch, M.F., Bocarsly, A.B. Probing the mechanism of aqueous CO_2 reduction on post-transition-metal electrodes using ATR-IR spectroelectrochemistry, ACS Catal. 2016, 6, 7824–7833.

[68] Solymosi, F. The bonding, structure and reactions of CO_2 adsorbed on clean and promoted metal surfaces, J. Mol. Catal. 1991, 65, 337–358.

[69] Koh, J.H., et al. Facile CO_2 electro-reduction to formate via oxygen bidentate intermediate stabilized by high-index planes of Bi dendrite catalyst, ACS Catal 2017, 7, 5071–5077.

[70] Feaster, J.T., et al. Understanding selectivity for the electrochemical reduction of carbon dioxide to formic acid and carbon monoxide on metal electrodes, ACS Catal 2017, 7, 4822–4827.

[71] Hansen, H.A., Varley, J.B., Peterson, A.A., Nørskov, J.K. Understanding trends in the electrocatalytic activity of metals and enzymes for CO_2 reduction to CO, J. Phys. Chem. Lett. 2013, 4, 388–392.

[72] Wang, L., et al. Electrochemical carbon monoxide reduction on polycrystalline copper: Effects of potential, pressure, and pH on selectivity toward multicarbon and oxygenated products, ACS Catal. 2018, 8, 7445–7454.

[73] Liu, X., et al. Understanding trends in electrochemical carbon dioxide reduction rates, Nat. Commun. 2017, 8, 15438.

[74] Peterson, A.A., Abild-Pedersen, F., Studt, F., Rossmeisl, J., Nørskov, J.K. How copper catalyzes the electroreduction of carbon dioxide into hydrocarbon fuels, Energy Environ. Sci 2010, 3, 1311–1315.

[75] Nie, X., Esopi, M.R., Janik, M.J., Asthagiri, A. Selectivity of CO_2 reduction on copper electrodes: The role of the kinetics of elementary steps, Angew. Chem. Int. Ed. 2013, 52, 2459–2462.

[76] Hatsukade, T., Kuhl, K.P., Cave, E.R., Abram, D.N., Jaramillo, T.F. Insights into the electrocatalytic reduction of CO_2 on metallic silver surfaces, Phys. Chem. Chem. Phys. 2014, 16, 13814–13819.

[77] Schouten, K.J.P., Kwon, Y., van der Ham, C.J.M., Qin, Z., Koper, M.T.M. A new mechanism for the selectivity to C_1 and C_2 species in the electrochemical reduction of carbon dioxide on copper electrodes, Chem. Sci. 2011, 2, 1902–1909.

[78] Li, J., et al. Enhanced multi-carbon alcohol electroproduction from CO via modulated hydrogen adsorption, Nat. Commun. 2020, 11, 3685.

[79] Pérez-Gallent, E., Figueiredo, M.C., Calle-Vallejo, F., Koper, M.T.M. Spectroscopic observation of a hydrogenated CO dimer intermediate during CO reduction on Cu(100) electrodes, Angew. Chem. Int. Ed. 2017, 56, 3621–3624.

[80] Schouten, K.J.P., Qin, Z., Pérez Gallent, E., Koper, M.T.M. Two pathways for the formation of ethylene in CO reduction on single-crystal copper electrodes, J. Am. Chem. Soc. 2012, 134, 9864–9867.

[81] Calle-Vallejo, F., Koper, M.T.M. Theoretical considerations on the electroreduction of CO to C_2 species on Cu(100) electrodes, Angew. Chem. Int. Ed. 2013, 52, 7282–7285.

[82] Lum, Y., Yue, B., Lobaccaro, P., Bell, A.T., Ager, J.W. Optimizing C–C coupling on oxide-derived copper catalysts for electrochemical CO_2 reduction, J. Phys. Chem. C 2017, 121, 14191–14203.

[83] Montoya, J.H., Shi, C., Chan, K., Nørskov, J.K. Theoretical insights into a CO dimerization mechanism in CO_2 electroreduction, J. Phys. Chem. Lett. 2015, 6, 2032–2037.

[84] Goodpaster, J.D., Bell, A.T., Head-Gordon, M. Identification of possible pathways for C–C bond formation during electrochemical reduction of CO_2: New theoretical insights from an improved electrochemical model, J. Phys. Chem. Lett. 2016, 7, 1471–1477.

[85] Garza, A.J., Bell, A.T., Head-Gordon, M. Mechanism of CO_2 reduction at copper surfaces: Pathways to C_2 products, ACS Catal. 2018, 8, 1490–1499.

[86] Luo, W., Nie, X., Janik, M.J., Asthagiri, A. Facet dependence of CO_2 reduction paths on Cu electrodes, ACS Catal. 2016, 6, 219–229.

[87] Cheng, T., Xiao, H., Goddard, W.A. Full atomistic reaction mechanism with kinetics for CO reduction on Cu(100) from ab initio molecular dynamics free-energy calculations at 298 K, Proc. Natl. Acad. Sci. 2017, 114, 1795–1800.

[88] Xiao, H., Cheng, T., Goddard, W.A. Atomistic mechanisms underlying selectivities in C_1 and C_2 products from electrochemical reduction of CO on Cu(111), J. Am. Chem. Soc. 2017, 139, 130–136.

[89] Birdja, Y.Y., et al. Advances and challenges in understanding the electrocatalytic conversion of carbon dioxide to fuels, Nat. Energy 2019, 4, 732–745.

[90] Kuhl, K.P., Cave, E.R., Abram, D.N., Jaramillo, T.F. New insights into the electrochemical reduction of carbon dioxide on metallic copper surfaces, Energy Environ. Sci 2012, 5, 7050–7059.

[91] Kortlever, R., et al. Palladium–gold catalyst for the electrochemical reduction of CO_2 to C_1–C_5 hydrocarbons, Chem. Commun. 2016, 52, 10229–10232.

[92] Koper, M.T.M. Thermodynamic theory of multi-electron transfer reactions: Implications for electrocatalysis, J. Electroanal. Chem. 2011, 660, 254–260.

[93] Calle-Vallejo, F., Koper, M.T.M. First-principles computational electrochemistry: Achievements and challenges, Electrochim. Acta 2012, 84, 3–11.

[94] Peterson, A.A., Nørskov, J.K. Activity descriptors for CO_2 electroreduction to methane on transition-metal catalysts, J. Phys. Chem. Lett. 2012, 3, 251–258.

[95] Halck, N.B., Petrykin, V., Krtil, P., Rossmeisl, J. Beyond the volcano limitations in electrocatalysis – oxygen evolution reaction, Phys. Chem. Chem. Phys. 2014, 16, 13682–13688.

[96] Xiong, L., et al. Breaking the linear scaling relationship by compositional and structural crafting of ternary Cu–Au/Ag nanoframes for electrocatalytic ethylene production, Angew. Chem. Int. Ed. 2021, 60, 2508–2518.

[97] Ross, M.B., et al. Designing materials for electrochemical carbon dioxide recycling, Nat. Catal. 2019, 2, 648–658.

[98] Miller, A.J.M., Labinger, J.A., Bercaw, J.E. Reductive coupling of carbon monoxide in a rhenium carbonyl complex with pendant Lewis acids, J. Am. Chem. Soc. 2008, 130, 11874–11875.

[99] Skúlason, E., et al. Modeling the electrochemical hydrogen oxidation and evolution reactions on the basis of density functional theory calculations, J. Phys. Chem. C 2010, 114, 18182–18197.

[100] Mariano, R.G., McKelvey, K., White, H.S., Kanan, M.W. Selective increase in CO_2 electroreduction activity at grain-boundary surface terminations, Science 2017, 358, 1187–1192.

[101] Verma, S., et al. Insights into the low overpotential electroreduction of CO_2 to CO on a supported gold catalyst in an alkaline flow electrolyzer, ACS Energy Lett 2018, 3, 193–198.

[102] Lee, H.-E., et al. Concave rhombic dodecahedral Au nanocatalyst with multiple high-index facets for CO_2 reduction, ACS Nano 2015, 9, 8384–8393.

[103] Zhu, W., et al. Active and selective conversion of CO_2 to CO on ultrathin Au nanowires, J. Am. Chem. Soc. 2014, 136, 16132–16135.

[104] Rosen, J., et al. Mechanistic insights into the electrochemical reduction of CO_2 to CO on nanostructured Ag surfaces, ACS Catal 2015, 5, 4293–4299.

[105] Ma, S., et al. Carbon nanotube containing Ag catalyst layers for efficient and selective reduction of carbon dioxide, J. Mater. Chem. A 2016, 4, 8573–8578.

[106] Kottakkat, T., et al. Electrodeposited AgCu foam catalysts for enhanced reduction of CO_2 to CO, ACS Appl. Mater. Interfaces 2019, 11, 14734–14744.

[107] Lee, H., Kim, J., Choi, I., Ahn, S.H. Nanostructured Ag/In/Cu foam catalyst for electrochemical reduction of CO_2 to CO, Electrochim. Acta 2019, 323, 133102.

[108] Gao, D., et al. Enhancing CO_2 electroreduction with the metal–oxide interface, J. Am. Chem. Soc. 2017, 139, 5652–5655.

[109] Kim, D., Resasco, J., Yu, Y., Asiri, A.M., Yang, P. Synergistic geometric and electronic effects for electrochemical reduction of carbon dioxide using gold–copper bimetallic nanoparticles, Nat. Commun. 2014, 5, 4948.

[110] Rosen, J., et al. Electrodeposited Zn dendrites with enhanced CO selectivity for electrocatalytic CO_2 reduction, ACS Catal 2015, 5, 4586–4591.

[111] Quan, F., Zhong, D., Song, H., Jia, F., Zhang, L. A highly efficient zinc catalyst for selective electroreduction of carbon dioxide in aqueous NaCl solution, J. Mater. Chem. A 2015, 3, 16409–16413.

[112] Li, Y.H., Liu, P.F., Li, C., Yang, H.G. Sharp-tipped zinc nanowires as an efficient electrocatalyst for carbon dioxide reduction, Chem. Eur. J. 2018, 24, 15486–15490.

[113] Yamamoto, T., Tryk, D.A., Hashimoto, K., Fujishima, A., Okawa, M. Electrochemical reduction of CO_2 in the micropores of activated carbon fibers, J. Electrochem. Soc. 2000, 147, 3393.

[114] Li, X., et al. Exclusive Ni–N_4 sites realize near-unity CO selectivity for electrochemical CO_2 reduction, J. Am. Chem. Soc. 2017, 139, 14889–14892.

[115] Nguyen, T.N., Salehi, M., Le, Q.V., Seifitokaldani, A., Dinh, C.T. Fundamentals of electrochemical CO_2 reduction on single-metal-atom catalysts, ACS Catal. 2020, 10, 10068–10095.

[116] Bi, W., et al. Surface immobilization of transition metal ions on nitrogen-doped graphene realizing high-efficient and selective CO_2 reduction, Adv. Mater. 2018, 30, 1706617.

[117] Cheng, Y., et al. Atomically dispersed transition metals on carbon nanotubes with ultrahigh loading for selective electrochemical carbon dioxide reduction, Adv. Mater. 2018, 30, 1706287.

[118] Yang, F., et al. Highly efficient CO_2 electroreduction on ZnN_4-based single-atom catalyst, Angew. Chem. Int. Ed. 2018, 57, 12303–12307.

[119] Zhang, E., et al. Bismuth single atoms resulting from transformation of metal–organic frameworks and their use as electrocatalysts for CO_2 reduction, J. Am. Chem. Soc. 2019, 141, 16569–16573.

[120] Sheng, W., et al. Electrochemical reduction of CO_2 to synthesis gas with controlled CO/H_2 ratios, Energy Environ. Sci. 2017, 10, 1180–1185.

[121] Lee, J.H., et al. Tuning the activity and selectivity of electroreduction of CO_2 to synthesis gas using bimetallic catalysts, Nat. Commun. 2019, 10, 3724.

[122] Qin, B., et al. Efficient electrochemical reduction of CO_2 into CO promoted by sulfur vacancies, Nano Energy 2019, 60, 43–51.

[123] Meng, N., Liu, C., Liu, Y., Yu, Y., Zhang, B. Efficient electrosynthesis of syngas with tunable CO/H_2 ratios over $Zn_xCd_{1-x}S$-amine inorganic–organic hybrids, Angew. Chem. Int. Ed. 2019, 58, 18908–18912.

[124] Xie, H., et al. Boosting tunable syngas formation via electrochemical CO_2 reduction on Cu/In_2O_3 core/shell nanoparticles, ACS Appl. Mater. Interfaces 2018, 10, 36996–37004.

[125] Mistry, H., et al. Exceptional size-dependent activity enhancement in the electroreduction of CO_2 over Au nanoparticles, J. Am. Chem. Soc. 2014, 136, 16473–16476.

[126] Jeon, H.S., et al. Operando evolution of the structure and oxidation state of size-controlled Zn nanoparticles during CO_2 electroreduction, J. Am. Chem. Soc. 2018, 140, 9383–9386.

[127] Marques Mota, F., et al. Toward an effective control of the H_2 to CO ratio of syngas through CO_2 electroreduction over immobilized gold nanoparticles on layered titanate nanosheets, ACS Catal. 2018, 8, 4364–4374.

[128] Qin, B., et al. Electrochemical reduction of CO_2 into tunable syngas production by regulating the crystal facets of earth-abundant Zn catalyst, ACS Appl. Mater. Interfaces 2018, 10, 20530–20539.

[129] Geng, Z., et al. Oxygen vacancies in ZnO nanosheets enhance CO_2 electrochemical reduction to CO, Angew. Chem. Int. Ed. 2018, 57, 6054–6059.

[130] Daiyan, R., et al. Electroreduction of CO_2 to CO on a mesoporous carbon catalyst with progressively removed nitrogen moieties, ACS Energy Lett. 2018, 3, 2292–2298.

[131] Lei, T., et al. Continuous electroreduction of carbon dioxide to formate on tin nanoelectrode using alkaline membrane cell configuration in aqueous medium, Catal. Today 2018, 318, 32–38.

[132] Li, F., Chen, L., Knowles, G.P., MacFarlane, D.R., Zhang, J. Hierarchical mesoporous SnO_2 nanosheets on carbon cloth: A robust and flexible electrocatalyst for CO_2 reduction with high efficiency and selectivity, Angew. Chem. Int. Ed. 2017, 56, 505–509.

[133] Luc, W., et al. Ag–Sn bimetallic catalyst with a core–shell structure for CO_2 reduction, J. Am. Chem. Soc. 2017, 139, 1885–1893.

[134] Lai, Q., Yang, N., Yuan, G. Highly efficient In–Sn alloy catalysts for electrochemical reduction of CO_2 to formate, Electrochem. Commun. 2017, 83, 24–27.

[135] Wen, G., et al. Orbital interactions in Bi-Sn bimetallic electrocatalysts for highly selective electrochemical CO_2 reduction toward formate production, Adv. Energy Mater. 2018, 8, 1802427.

[136] Peng, L., et al. Self-growing Cu/Sn bimetallic electrocatalysts on nitrogen-doped porous carbon cloth with 3D-hierarchical honeycomb structure for highly active carbon dioxide reduction, Appl. Catal. B 2020, 264, 118447.

[137] Choi, S.Y., Jeong, S.K., Kim, H.J., Baek, I.-H., Park, K.T. Electrochemical reduction of carbon dioxide to formate on tin–lead alloys, ACS Sustain. Chem. Eng. 2016, 4, 1311–1318.

[138] Bai, X., et al. Exclusive formation of formic acid from CO_2 electroreduction by a tunable Pd-Sn alloy, Angew. Chem. Int. Ed. 2017, 56, 12219–12223.

[139] Ye, K., et al. Synergy effects on Sn-Cu alloy catalyst for efficient CO_2 electroreduction to formate with high mass activity, Sci. Bull. 2020, 65, 711–719.

[140] Pander, J.E., Lum, J.W.J., Yeo, B.S. The importance of morphology on the activity of lead cathodes for the reduction of carbon dioxide to formate, J. Mater. Chem. A 2019, 7, 4093–4101.

[141] Zhang, H., et al. Selective electro-reduction of CO_2 to formate on nanostructured Bi from reduction of BiOCl nanosheets, Electrochem. Commun. 2014, 46, 63–66.

[142] Li, C., et al. Carbon dioxide photo/electroreduction with cobalt, J. Mater. Chem. A 2019, 7, 16622–16642.

[143] Gao, S., et al. Atomic layer confined vacancies for atomic-level insights into carbon dioxide electroreduction, Nat. Commun. 2017, 8, 14503.

[144] Gao, S., et al. Partially oxidized atomic cobalt layers for carbon dioxide electroreduction to liquid fuel, Nature 2016, 529, 68–71.

[145] Zu, X., et al. Efficient and robust carbon dioxide electroreduction enabled by atomically dispersed $Sn^{\delta+}$ sites, Adv. Mater 2019, 31, 1808135.

[146] Taheri, A., Thompson, E.J., Fettinger, J.C., Berben, L.A. An iron electrocatalyst for selective reduction of CO_2 to formate in water: Including thermochemical insights, ACS Catal. 2015, 5, 7140–7151.

[147] Ma, W., et al. Promoting electrocatalytic CO_2 reduction to formate via sulfur-boosting water activation on indium surfaces, Nat. Commun. 2019, 10, 892.

[148] Huang, Y., Deng, Y., Handoko, A.D., Goh, G.K.L., Yeo, B.S. Rational design of sulfur-doped copper catalysts for the selective electroreduction of carbon dioxide to formate, ChemSusChem 2018, 11, 320–326.

[149] Zheng, X., et al. Sulfur-modulated tin sites enable highly selective electrochemical reduction of CO_2 to formate, Joule 2017, 1, 794–805.

[150] Kopljar, D., Inan, A., Vindayer, P., Wagner, N., Klemm, E. Electrochemical reduction of CO_2 to formate at high current density using gas diffusion electrodes, J. Appl. Electrochem. 2014, 44, 1107–1116.

[151] Sen, S., Brown, S.M., Leonard, M., Brushett, F.R. Electroreduction of carbon dioxide to formate at high current densities using tin and tin oxide gas diffusion electrodes, J. Appl. Electrochem. 2019, 49, 917–928.

[152] Baruch, M.F., Pander, J.E., White, J.L., Bocarsly, A.B. Mechanistic insights into the reduction of CO_2 on tin electrodes using in situ ATR-IR spectroscopy, ACS Catal. 2015, 5, 3148–3156.

[153] Damas, G.B., et al. On the mechanism of carbon dioxide reduction on Sn-based electrodes: Insights into the role of oxide surfaces, Catalysts 2019, 9, 636.

[154] Qian, Y., Liu, Y., Tang, H., Lin, B.-L. Highly efficient electroreduction of CO_2 to formate by nanorod@2d nanosheets SnO, J. CO2 Util. 2020, 42, 101287.

[155] Sun, Z., Ma, T., Tao, H., Fan, Q., Han, B. Fundamentals and challenges of electrochemical CO_2 reduction using two-dimensional materials, Chem 2017, 3, 560–587.

[156] Hori, Y., Wakebe, H., Tsukamoto, T., Koga, O. Adsorption of CO accompanied with simultaneous charge transfer on copper single crystal electrodes related with electrochemical reduction of CO_2 to hydrocarbons, Surf. Sci. 1995, 335, 258–263.

[157] Ou, L., Chen, Y., Jin, J. The origin of CO_2 electroreduction into C_1 and C_2 species: Mechanistic understanding on the product selectivity of Cu single-crystal faces, Chem. Phys. Lett. 2018, 710, 175–179.

[158] Wang, Z., Yang, G., Zhang, Z., Jin, M., Yin, Y. Selectivity on etching: Creation of high-energy facets on copper nanocrystals for CO_2 electrochemical reduction, ACS Nano 2016, 10, 4559–4564.

[159] Manthiram, K., Beberwyck, B.J., Alivisatos, A.P. Enhanced electrochemical methanation of carbon dioxide with a dispersible nanoscale copper catalyst, J. Am. Chem. Soc. 2014, 136, 13319–13325.

[160] Li, Y., et al. Structure-sensitive CO_2 electroreduction to hydrocarbons on ultrathin 5-fold twinned copper nanowires, Nano Lett. 2017, 17, 1312–1317.

[161] Le, M., et al. Electrochemical reduction of CO_2 to CH_3OH at copper oxide surfaces, J. Electrochem. Soc. 2011, 158, E45.

[162] Azenha, C., Mateos-Pedrero, C., Alvarez-Guerra, M., Irabien, A., Mendes, A. Enhancement of the electrochemical reduction of CO_2 to methanol and suppression of H_2 evolution over CuO nanowires, Electrochim. Acta 2020, 363, 137207.

[163] Zhao, K., Liu, Y., Quan, X., Chen, S., Yu, H. CO_2 electroreduction at low overpotential on oxide-derived Cu/carbons fabricated from metal organic framework, ACS Appl. Mater. Interfaces 2017, 9, 5302–5311.

[164] Wang, Y., et al. Single-atomic Cu with multiple oxygen vacancies on ceria for electrocatalytic CO_2 reduction to CH_4, ACS Catal. 2018, 8, 7113–7119.

[165] Yang, H., et al. Scalable production of efficient single-atom copper decorated carbon membranes for CO_2 electroreduction to methanol, J. Am. Chem. Soc. 2019, 141, 12717–12723.

[166] Tan, X., et al. Restructuring of Cu_2O to Cu_2O@Cu-metal–organic frameworks for selective electrochemical reduction of CO_2, ACS Appl. Mater. Interfaces 2019, 11, 9904–9910.

[167] Takatsuji, Y., et al. Highly selective methane production through electrochemical CO_2 reduction by electrolytically plated Cu-Co electrode, Electrocatalysis 2019, 10, 29–34.

[168] Zhou, B., et al. Highly efficient binary copper–iron catalyst for photoelectrochemical carbon dioxide reduction toward methane, Proc, Natl. Acad. Sci. 2020, 117, 1330–1338.

[169] Wang, Z., et al. Highly selective electrocatalytic reduction of CO_2 into methane on Cu–Bi nanoalloys, J. Phys. Chem. Lett. 2020, 11, 7261–7266.

[170] Jia, F., Yu, X., Zhang, L. Enhanced selectivity for the electrochemical reduction of CO_2 to alcohols in aqueous solution with nanostructured Cu–Au alloy as catalyst, J. Power Sources 2014, 252, 85–89.

[171] Hori, Y., Murata, A., Takahashi, R., Suzuki, S. Electroreduction of carbon monoxide to methane and ethylene at a copper electrode in aqueous solutions at ambient temperature and pressure, J. Am. Chem. Soc. 1987, 109, 5022–5023.

[172] Hahn, C., et al. Engineering Cu surfaces for the electrocatalytic conversion of CO_2: Controlling selectivity toward oxygenates and hydrocarbons, Proc. Natl. Acad. Sci. 2017, 114, 5918–5923.

[173] Hori, Y., Takahashi, I., Koga, O., Hoshi, N. Electrochemical reduction of carbon dioxide at various series of copper single crystal electrodes, J. Mol. Catal. A: Chem. 2003, 199, 39–47.

[174] Tang, W., et al. The importance of surface morphology in controlling the selectivity of polycrystalline copper for CO_2 electroreduction, Phys. Chem. Chem. Phys. 2012, 14, 76–81.

[175] Hori, Y., Takahashi, I., Koga, O., Hoshi, N. Selective formation of C_2 compounds from electrochemical reduction of CO_2 at a series of copper single crystal electrodes, J. Phys. Chem. B 2002, 106, 15–17.

[176] Loiudice, A., et al. Tailoring copper nanocrystals towards C_2 products in electrochemical CO_2 reduction, Angew. Chem. Int. Ed. 2016, 55, 5789–5792.

[177] Wang, Y., et al. Catalyst synthesis under CO_2 electroreduction favours faceting and promotes renewable fuels electrosynthesis, Nat. Catal. 2020, 3, 98–106.

[178] Reller, C., et al. Selective electroreduction of CO_2 toward ethylene on nano dendritic copper catalysts at high current density, Adv. Energy Mater. 2017, 7, 1602114.

[179] Hoang, T.T.H., Ma, S., Gold, J.I., Kenis, P.J.A., Gewirth, A.A. Nanoporous copper films by additive-controlled electrodeposition: CO_2 reduction catalysis, ACS Catal. 2017, 7, 3313–3321.

[180] Zhuang, -T.-T., et al. Steering post-C–C coupling selectivity enables high efficiency electroreduction of carbon dioxide to multi-carbon alcohols, Nat. Catal. 2018, 1, 421–428.

[181] Kwon, Y., Lum, Y., Clark, E.L., Ager, J.W., Bell, A.T. CO_2 electroreduction with enhanced ethylene and ethanol selectivity by nanostructuring polycrystalline copper, ChemElectroChem. 2016, 3, 1012–1019.

[182] Möller, T., et al. Electrocatalytic CO_2 reduction on CuO_x nanocubes: Tracking the evolution of chemical state, geometric structure, and catalytic selectivity using operando spectroscopy, Angew. Chem. Int. Ed. 2020, 59, 17974–17983.

[183] Ren, D., et al. Selective electrochemical reduction of carbon dioxide to ethylene and ethanol on copper(I) oxide catalysts, ACS Catal. 2015, 5, 2814–2821.

[184] Mistry, H., et al. Highly selective plasma-activated copper catalysts for carbon dioxide reduction to ethylene, Nat. Commun. 2016, 7, 12123.

[185] Xiao, H., Goddard, W.A., Cheng, T., Liu, Y. Cu metal embedded in oxidized matrix catalyst to promote CO_2 activation and CO dimerization for electrochemical reduction of CO_2, Proc. Natl. Acad. Sci. 2017, 114, 6685–6688.

[186] Liang, Z.-Q., et al. Copper-on-nitride enhances the stable electrosynthesis of multi-carbon products from CO_2, Nat. Commun. 2018, 9, 3828.

[187] Yano, H., Tanaka, T., Nakayama, M., Ogura, K. Selective electrochemical reduction of CO_2 to ethylene at a three-phase interface on copper(I) halide-confined Cu-mesh electrodes in acidic solutions of potassium halides, J. Electroanal. Chem. 2004, 565, 287–293.

[188] Ma, W., et al. Electrocatalytic reduction of CO_2 to ethylene and ethanol through hydrogen-assisted C–C coupling over fluorine-modified copper, Nat. Catal. 2020, 3, 478–487.

[189] Ma, S., et al. Electroreduction of carbon dioxide to hydrocarbons using bimetallic Cu–Pd catalysts with different mixing patterns, J. Am. Chem. Soc. 2017, 139, 47–50.

[190] Chen, C.S., Wan, J.H., Yeo, B.S. Electrochemical reduction of carbon dioxide to ethane using nanostructured Cu_2O-derived copper catalyst and palladium(II) chloride, J. Phys. Chem. C 2015, 119, 26875–26882.

[191] Ren, D., Ang, B.S.-H., Yeo, B.S. Tuning the selectivity of carbon dioxide electroreduction toward ethanol on oxide-derived Cu_xZn catalysts, ACS Catal. 2016, 6, 8239–8247.

[192] Li, Y.C., et al. Binding site diversity promotes CO_2 electroreduction to ethanol, J. Am. Chem. Soc. 2019, 141, 8584–8591.

[193] Zhu, Q., et al. Carbon dioxide electroreduction to C_2 products over copper-cuprous oxide derived from electrosynthesized copper complex, Nat. Commun. 2019, 10, 3851.

[194] Genovese, C., Ampelli, C., Perathoner, S., Centi, G. Mechanism of C–C bond formation in the electrocatalytic reduction of CO_2 to acetic acid. A challenging reaction to use renewable energy with chemistry, Green Chem. 2017, 19, 2406–2415.

[195] Wang, H., et al. Synergistic enhancement of electrocatalytic CO_2 reduction to C_2 oxygenates at nitrogen-doped nanodiamonds/Cu interface, Nat. Nanotechnol. 2020, 15, 131–137.

[196] Zhuang, -T.-T., et al. Copper nanocavities confine intermediates for efficient electrosynthesis of C_3 alcohol fuels from carbon monoxide, Nat. Catal. 2018, 1, 946–951.

[197] Kim, D., Kley, C.S., Li, Y., Yang, P. Copper nanoparticle ensembles for selective electroreduction of CO_2 to C_2–C_3 products, Proc. Natl. Acad. Sci. 2017, 114, 10560–10565.

[198] Lee, S., Kim, D., Lee, J. Electrocatalytic production of C_3-C_4 compounds by conversion of CO_2 on a chloride-induced Bi-phasic Cu_2O-Cu catalyst, Angew. Chem. Int. Ed. 2015, 54, 14701–14705.

[199] Ren, D., Wong, N.T., Handoko, A.D., Huang, Y., Yeo, B.S. Mechanistic insights into the enhanced activity and stability of agglomerated Cu nanocrystals for the electrochemical reduction of carbon dioxide to n-propanol, J. Phys. Chem. Lett. 2016, 7, 20–24.

[200] Kim, T., et al. Enhancing C_2–C_3 production from CO_2 on copper electrocatalysts via a potential-dependent mesostructure, ACS Appl. Energy Mater. 2018, 1, 1965–1972.

[201] Romero Cuellar, N.S., Wiesner-Fleischer, K., Fleischer, M., Rucki, A., Hinrichsen, O. Advantages of CO over CO_2 as reactant for electrochemical reduction to ethylene, ethanol and n-propanol on gas diffusion electrodes at high current densities, Electrochim. Acta 2019, 307, 164–175.

[202] Romero Cuellar, N.S., et al. Two-step electrochemical reduction of CO_2 towards multi-carbon products at high current densities, J. CO2 Util. 2020, 36, 263–275.

[203] Huang, J., Mensi, M., Oveisi, E., Mantella, V., Buonsanti, R. Structural sensitivities in bimetallic catalysts for electrochemical CO_2 reduction revealed by Ag–Cu nanodimers, J. Am. Chem. Soc. 2019, 141, 2490–2499.

[204] O'Mara, P.B., et al. Cascade reactions in nanozymes: Spatially separated active sites inside Ag-core–porous-Cu-shell nanoparticles for multistep carbon dioxide reduction to higher organic molecules, J. Am. Chem. Soc. 2019, 141, 14093–14097.

[205] Hoang, T.T.H., et al. Nanoporous copper–silver alloys by additive-controlled electrodeposition for the selective electroreduction of CO_2 to ethylene and ethanol, J. Am. Chem. Soc. 2018, 140, 5791–5797.

[206] Feng, Y., et al. Laser-prepared CuZn alloy catalyst for selective electrochemical reduction of CO_2 to ethylene, Langmuir 2018, 34, 13544–13549.

[207] Li, Z., et al. CuO/ZnO/C electrocatalysts for CO_2-to-C_{2+} products conversion with high yield: On the effect of geometric structure and composition, Appl. Catal. A 2020, 606, 117829.

[208] Zhang, H., et al. Computational and experimental demonstrations of one-pot tandem catalysis for electrochemical carbon dioxide reduction to methane, Nat. Commun. 2019, 10, 3340.

[209] Zhang, S., et al. Polymer-supported cupd nanoalloy as a synergistic catalyst for electrocatalytic reduction of carbon dioxide to methane, Proc. Natl. Acad. Sci. 2015, 112, 15809–15814.

[210] Guo, X., et al. Composition dependent activity of Cu–Pt nanocrystals for electrochemical reduction of CO_2, Chem. Commun. 2015, 51, 1345–1348.

[211] Kuhl, K.P., et al. Electrocatalytic conversion of carbon dioxide to methane and methanol on transition metal surfaces, J. Am. Chem. Soc. 2014, 136, 14107–14113.

[212] Wu, Y., Jiang, Z., Lu, X., Liang, Y., Wang, H. Domino electroreduction of CO_2 to methanol on a molecular catalyst, Nature 2019, 575, 639–642.

[213] Huang, J., Hu, Q., Guo, X., Zeng, Q., Wang, L. Rethinking $Co(CO_3)_{0.5}(OH)\cdot0.11H_2O$: A new property for highly selective electrochemical reduction of carbon dioxide to methanol in aqueous solution, Green Chem. 2018, 20, 2967–2972.

[214] Zhang, W., et al. Electrochemical reduction of carbon dioxide to methanol on hierarchical Pd/SnO_2 nanosheets with abundant Pd–O–Sn interfaces, Angew. Chem. Int. Ed. 2018, 57, 9475–9479.

[215] Umeda, M., Niitsuma, Y., Horikawa, T., Matsuda, S., Osawa, M. Electrochemical reduction of CO_2 to methane on platinum catalysts without overpotentials: Strategies for improving conversion efficiency, ACS Appl. Energy Mater. 2020, 3, 1119–1127.

[216] Qu, J., Zhang, X., Wang, Y., Xie, C. Electrochemical reduction of CO_2 on RuO_2/TiO_2 nanotubes composite modified Pt electrode, Electrochim. Acta 2005, 50, 3576–3580.

[217] Han, L., et al. Stable and efficient single-atom Zn catalyst for CO_2 reduction to CH_4, J. Am. Chem. Soc. 2020, 142, 12563–12567.

[218] Gonglach, S., et al. Molecular cobalt corrole complex for the heterogeneous electrocatalytic reduction of carbon dioxide, Nat. Commun. 2019, 10, 3864.

[219] De, R., et al. Electrocatalytic reduction of CO_2 to acetic acid by a molecular manganese corrole complex, Angew. Chem. Int. Ed. 2020, 59, 10527–10534.

[220] Liu, Y., et al. Steering CO_2 electroreduction toward ethanol production by a surface-bound Ru polypyridyl carbene catalyst on N-doped porous carbon, Proc. Natl. Acad. Sci. 2019, 116, 26353–26358.

[221] Liu, Y., Chen, S., Quan, X., Yu, H. Efficient electrochemical reduction of carbon dioxide to acetate on nitrogen-doped nanodiamond, J. Am. Chem. Soc. 2015, 137, 11631–11636.

[222] Stamenkovic, V.R., Strmcnik, D., Lopes, P.P., Markovic, N.M. Energy and fuels from electrochemical interfaces, Nat. Mater. 2017, 16, 57–69.

[223] Sheng, W., Myint, M., Chen, J.G., Yan, Y. Correlating the hydrogen evolution reaction activity in alkaline electrolytes with the hydrogen binding energy on monometallic surfaces, Energy Environ. Sci. 2013, 6, 1509–1512.

[224] Durst, J., et al. New insights into the electrochemical hydrogen oxidation and evolution reaction mechanism, Energy Environ. Sci. 2014, 7, 2255–2260.

[225] Dinh, C.-T., et al. CO_2 electroreduction to ethylene via hydroxide-mediated copper catalysis at an abrupt interface, Science. 2018, 360, 783–787.

[226] Gupta, N., Gattrell, M., MacDougall, B. Calculation for the cathode surface concentrations in the electrochemical reduction of CO_2 in $KHCO_3$ solutions, J. Appl. Electrochem. 2006, 36, 161–172.

[227] Hori, Y., Murata, A., Takahashi, R. Formation of hydrocarbons in the electrochemical reduction of carbon dioxide at a copper electrode in aqueous solution, J. Chem. Soc., Faraday Trans. 1 1989, 85, 2309–2326.

[228] Varela, A.S., Kroschel, M., Reier, T., Strasser, P. Controlling the selectivity of CO_2 electroreduction on copper: The effect of the electrolyte concentration and the importance of the local pH, Catal. Today 2016, 260, 8–13.

[229] Resasco, J., Lum, Y., Clark, E., Zeledon, J.Z., Bell, A.T. Effects of anion identity and concentration on electrochemical reduction of CO_2, ChemElectroChem 2018, 5, 1064–1072.

[230] Montoya, J.H., Peterson, A.A., Nørskov, J.K. Insights into C-C coupling in CO_2 electroreduction on copper electrodes, ChemCatChem 2013, 5, 737–742.

[231] Schouten, K.J.P., Pérez Gallent, E., Koper, M.T.M. The influence of pH on the reduction of CO and CO_2 to hydrocarbons on copper electrodes, J. Electroanal. Chem. 2014, 716, 53–57.

[232] Gattrell, M., Gupta, N., Co, A. A review of the aqueous electrochemical reduction of CO_2 to hydrocarbons at copper, J. Electroanal. Chem. 2006, 594, 1–19.

[233] Wuttig, A., Yoon, Y., Ryu, J., Surendranath, Y. Bicarbonate is not a general acid in Au-catalyzed CO_2 electroreduction, J. Am. Chem. Soc. 2017, 139, 17109–17113.

[234] Chen, Y., Li, C.W., Kanan, M.W. Aqueous CO_2 reduction at very low overpotential on oxide-derived Au nanoparticles, J. Am. Chem. Soc. 2012, 134, 19969–19972.

[235] Dunwell, M., et al. The central role of bicarbonate in the electrochemical reduction of carbon dioxide on gold, J. Am. Chem. Soc. 2017, 139, 3774–3783.

[236] Li, T., et al. Electrolytic conversion of bicarbonate into CO in a flow cell, Joule. 2019, 3, 1487–1497.

[237] Sreekanth, N., Phani, K.L. Selective reduction of CO_2 to formate through bicarbonate reduction on metal electrodes: New insights gained from SG/TC mode of SECM, Chem. Commun. 2014, 50, 11143–11146.

[238] Katsounaros, I., et al. The impact of spectator species on the interaction of H_2O_2 with platinum – implications for the oxygen reduction reaction pathways, Phys. Chem. Chem. Phys. 2013, 15, 8058–8068.

[239] Ogura, K., Ferrell, J.R., Cugini, A.V., Smotkin, E.S., Salazar-Villalpando, M.D. CO_2 attraction by specifically adsorbed anions and subsequent accelerated electrochemical reduction, Electrochim. Acta 2010, 56, 381–386.

[240] Ogura, K., Yano, H., Shirai, F. Catalytic reduction of CO_2 to ethylene by electrolysis at a three-phase interface, J. Electrochem. Soc. 2003, 150, D163.

[241] Bard, A.J. Inner-sphere heterogeneous electrode reactions. Electrocatalysis and photocatalysis: The challenge, J. Am. Chem. Soc. 2010, 132, 7559–7567.

[242] Ogura, K. Electrochemical reduction of carbon dioxide to ethylene: Mechanistic approach, J. CO2 Util. 2013, 1, 43–49.

[243] Varela, A.S., Ju, W., Reier, T., Strasser, P. Tuning the catalytic activity and selectivity of Cu for CO_2 electroreduction in the presence of halides, ACS Catal. 2016, 6, 2136–2144.
[244] Akira, M., Yoshio, H. Product selectivity affected by cationic species in electrochemical reduction of CO_2 and CO at a Cu electrode, Bull. Chem. Soc. Jpn. 1991, 64, 123–127.
[245] Paik, W., Andersen, T.N., Eyring, H. Kinetic studies of the electrolytic reduction of carbon dioxide on the mercury electrode, Electrochim. Acta 1969, 14, 1217–1232.
[246] Thorson, M.R., Siil, K.I., Kenis, P.J.A. Effect of cations on the electrochemical conversion of CO_2 to CO, J. Electrochem. Soc. 2012, 160, F69-F74.
[247] Strmcnik, D., et al. The role of non-covalent interactions in electrocatalytic fuel-cell reactions on platinum, Nat. Chem. 2009, 1, 466–472.
[248] Mills, J.N., McCrum, I.T., Janik, M.J. Alkali cation specific adsorption onto fcc(111) transition metal electrodes, Phys. Chem. Chem. Phys. 2014, 16, 13699–13707.
[249] Resasco, J., et al. Promoter effects of alkali metal cations on the electrochemical reduction of carbon dioxide, J. Am. Chem. Soc. 2017, 139, 11277–11287.
[250] Pérez-Gallent, E., Marcandalli, G., Figueiredo, M.C., Calle-Vallejo, F., Koper, M.T.M. Structure- and potential-dependent cation effects on CO reduction at copper single-crystal electrodes, J. Am. Chem. Soc. 2017, 139, 16412–16419.
[251] Sedighian Rasouli, A., et al. CO_2 electroreduction to methane at production rates exceeding 100 mA/cm^2, ACS Sustain. Chem. Eng. 2020, 8, 14668–14673.
[252] Singh, M.R., Kwon, Y., Lum, Y., Ager, J.W., Bell, A.T. Hydrolysis of electrolyte cations enhances the electrochemical reduction of CO_2 over Ag and Cu, J. Am. Chem. Soc. 2016, 138, 13006–13012.
[253] Gennaro, A., Isse, A.A., Vianello, E. Solubility and electrochemical determination of CO_2 in some dipolar aprotic solvents, J. Electroanal. Chem. Interf. Electrochem. 1990, 289, 203–215.
[254] Fogg, P.G.T. Carbon dioxide in non-aqueous solvents at pressures less than 200 kpa, Pergamon Press, Oxford, UK., 1992.
[255] Hansen, C.M. Hansen solubility parameters: A user's handbook, second edition, CRC Press, Boca Raton, FL, USA, R, 2007.
[256] Izutsu, K. Properties of solvents and solvent classification, Electrochemistry in nonaqueous solutions, John Wiley and Sons, Weinheim, Germany, 2002, 1–24.
[257] König, M., Vaes, J., Klemm, E., Pant, D. Solvents and supporting electrolytes in the electrocatalytic reduction of CO_2, iScience 2019, 19, 135–160.
[258] Amatore, C., Saveant, J.M. Mechanism and kinetic characteristics of the electrochemical reduction of carbon dioxide in media of low proton availability, J. Am. Chem. Soc. 1981, 103, 5021–5023.
[259] Kaiser, U., Heitz, E. On the mechanism of the electrochemical dimerization of CO_2 to oxalic acid, Ber. Bunsenges. Phys. Chem. 1973, 77, 818–823.
[260] Berto, T.C., Zhang, L., Hamers, R.J., Berry, J.F. Electrolyte dependence of CO_2 electroreduction: Tetraalkylammonium ions are not electrocatalysts, ACS Catal. 2015, 5, 703–707.
[261] Shi, J., et al. Electrochemical reduction of CO_2 into CO in tetrabutylammonium perchlorate/propylene carbonate: Water effects and mechanism, Electrochim. Acta 2017, 240, 114–121.
[262] Tomita, Y., Teruya, S., Koga, O., Hori, Y. Electrochemical reduction of carbon dioxide at a platinum electrode in acetonitrile-water mixtures, J. Electrochem. Soc. 2000, 147, 4164.
[263] Saeki, T., Hashimoto, K., Kimura, N., Omata, K., Fujishima, A. Electrochemical reduction of CO_2 with high current density in a CO_2 + methanol medium II. CO formation promoted by tetrabutylammonium cation, J. Electroanal. Chem. 1995, 390, 77–82.
[264] Setterfield-Price, B.M., Dryfe, R.A.W. The influence of electrolyte identity upon the electro-reduction of CO_2, J. Electroanal. Chem. 2014, 730, 48–58.

[265] Kai, T., Zhou, M., Duan, Z., Henkelman, G.A., Bard, A.J. Detection of CO_2^- in the electrochemical reduction of carbon dioxide in *N, N*-dimethylformamide by scanning electrochemical microscopy, J. Am. Chem. Soc. 2017, 139, 18552–18557.

[266] Figueiredo, M.C., Ledezma-Yanez, I., Koper, M.T.M. In situ spectroscopic study of CO_2 electroreduction at copper electrodes in acetonitrile, ACS Catal. 2016, 6, 2382–2392.

[267] Costentin, C., Robert, M., Savéant, J.-M. Catalysis of the electrochemical reduction of carbon dioxide, Chem. Soc. Rev. 2013, 42, 2423–2436.

[268] Aljabour, A., et al. Nanofibrous cobalt oxide for electrocatalysis of CO_2 reduction to carbon monoxide and formate in an acetonitrile-water electrolyte solution, Appl. Catal. B 2018, 229, 163–170.

[269] Paris, A.R., Bocarsly, A.B. High-efficiency conversion of CO_2 to oxalate in water is possible using a Cr-Ga oxide electrocatalyst, ACS Catal. 2019, 9, 2324–2333.

[270] Rudnev, A.V., et al. The promoting effect of water on the electroreduction of CO_2 in acetonitrile, Electrochim. Acta 2016, 189, 38–44.

[271] Ramesha, G.K., Brennecke, J.F., Kamat, P.V. Origin of catalytic effect in the reduction of CO_2 at nanostructured TiO_2 films, ACS Catal. 2014, 4, 3249–3254.

[272] Subramanian, S., Athira, K.R., Anbu Kulandainathan, M., Senthil Kumar, S., Barik, R.C. New insights into the electrochemical conversion of CO_2 to oxalate at stainless steel 304l cathode, J. CO2 Util. 2020, 36, 105–115.

[273] Kaneco, S., Iiba, K., Katsumata, H., Suzuki, T., Ohta, K. Electrochemical reduction of high pressure carbon dioxide at a Cu electrode in cold methanol with CsOH supporting salt, Chem. Eng. J. 2007, 128, 47–50.

[274] Saeki, T., Hashimoto, K., Kimura, N., Omata, K., Fujishima, A. Electrochemical reduction of CO_2 with high current density in a CO_2 + methanol medium at various metal electrodes, J. Electroanal. Chem. 1996, 404, 299–302.

[275] Kaneco, S., Iiba, K., Ohta, K., Mizuno, T., Saji, A. Electrochemical reduction of CO_2 on Au in KOH + methanol at low temperature, J. Electroanal. Chem. 1998, 441, 215–220.

[276] Uemoto, N., Furukawa, M., Tateishi, I., Katsumata, H., Kaneco, S. Electrochemical carbon dioxide reduction in methanol at Cu and Cu_2O-deposited carbon black electrodes, ChemEngineering 2019, 3, 15.

[277] Kaneco, S., Iiba, K., Katsumata, H., Suzuki, T., Ohta, K. Effect of sodium cation on the electrochemical reduction of CO_2 at a copper electrode in methanol, J. Solid State Electrochem. 2007, 11, 490–495.

[278] Kaneco, S., et al. Electrochemical reduction of carbon dioxide to ethylene with high faradaic efficiency at a Cu electrode in CsOH/methanol, Electrochim. Acta 1999, 44, 4701–4706.

[279] Kaneco, S., Katsumata, H., Suzuki, T., Ohta, K. Electrochemical reduction of CO_2 to methane at the Cu electrode in methanol with sodium supporting salts and its comparison with other alkaline salts, Energy Fuels 2006, 20, 409–414.

[280] Yang, D., Zhu, Q., Han, B. Electroreduction of CO_2 in ionic liquid-based electrolytes, Innovation. 2020, 1, 100016.

[281] Aghaie, M., Rezaei, N., Zendehboudi, S. A systematic review on CO_2 capture with ionic liquids: Current status and future prospects, Renew. Sustain. Energy Rev. 2018, 96, 502–525.

[282] Ramdin, M., de Loos, T.W., Vlugt, T.J.H. State-of-the-art of CO_2 capture with ionic liquids, Ind. Eng. Chem. Res. 2012, 51, 8149–8177.

[283] Shukla, S.K., Khokarale, S.G., Bui, T.Q., Mikkola, J.-P.T. Ionic liquids: Potential materials for carbon dioxide capture and utilization, Front. Mater. 2019, 6, 42.

[284] Bates, E.D., Mayton, R.D., Ntai, I., Davis, J.H. CO_2 capture by a task-specific ionic liquid, J. Am. Chem. Soc. 2002, 124, 926–927.

[285] Gurau, G., et al. Demonstration of chemisorption of carbon dioxide in 1,3-dialkylimidazolium acetate ionic liquids, Angew. Chem. Int. Ed. 2011, 50, 12024–12026.

[286] Rosen, B.A., et al. Ionic liquid–mediated selective conversion of CO_2 to CO at low overpotentials, Science 2011, 334, 643–644.

[287] Rosen, B.A., et al. In situ spectroscopic examination of a low overpotential pathway for carbon dioxide conversion to carbon monoxide, J. Phys. Chem. C 2012, 116, 15307–15312.

[288] Hanc-Scherer, F.A., Montiel, M.A., Montiel, V., Herrero, E., Sánchez-Sánchez, C.M. Surface structured platinum electrodes for the electrochemical reduction of carbon dioxide in imidazolium based ionic liquids, Phys. Chem. Chem. Phys. 2015, 17, 23909–23916.

[289] Zhu, W., et al. Monodisperse Au nanoparticles for selective electrocatalytic reduction of CO_2 to CO, J. Am. Chem. Soc. 2013, 135, 16833–16836.

[290] Yang, D.-W., et al. Electrochemical impedance studies of CO_2 reduction in ionic liquid/organic solvent electrolyte on Au electrode, Electrochim. Acta 2016, 189, 32–37.

[291] Zhao, S.-F., Horne, M., Bond, A.M., Zhang, J. Is the imidazolium cation a unique promoter for electrocatalytic reduction of carbon dioxide?, J. Phys. Chem. C 2016, 120, 23989–24001.

[292] Lau, G.P.S., et al. New insights into the role of imidazolium-based promoters for the electroreduction of CO_2 on a silver electrode, J. Am. Chem. Soc. 2016, 138, 7820–7823.

[293] Sun, L., Ramesha, G.K., Kamat, P.V., Brennecke, J.F. Switching the reaction course of electrochemical CO_2 reduction with ionic liquids, Langmuir 2014, 30, 6302–6308.

[294] Atifi, A., Boyce, D.W., DiMeglio, J.L., Rosenthal, J. Directing the outcome of CO_2 reduction at bismuth cathodes using varied ionic liquid promoters, ACS Catal. 2018, 8, 2857–2863.

[295] Rosen, B.A., Zhu, W., Kaul, G., Salehi-Khojin, A., Masel, R.I. Water enhancement of CO_2 conversion on silver in 1-ethyl-3-methylimidazolium tetrafluoroborate, J. Electrochem. Soc. 2012, 160, H138-H141.

[296] Rudnev, A.V., et al. Transport matters: Boosting CO_2 electroreduction in mixtures of [BMIM][BF_4]/water by enhanced diffusion, ChemPhysChem 2017, 18, 3153–3162.

[297] Krupiczka, R., Rotkegel, A., Ziobrowski, Z. Comparative study of CO_2 absorption in packed column using imidazolium based ionic liquids and MEA solution, Sep. Purif. Technol. 2015, 149, 228–236.

[298] Salehi-Khojin, A., et al. Nanoparticle silver catalysts that show enhanced activity for carbon dioxide electrolysis, J. Phys. Chem. C 2013, 117, 1627–1632.

[299] Medina-Ramos, J., Pupillo, R.C., Keane, T.P., DiMeglio, J.L., Rosenthal, J. Efficient conversion of CO_2 to CO using tin and other inexpensive and easily prepared post-transition metal catalysts, J. Am. Chem. Soc. 2015, 137, 5021–5027.

[300] DiMeglio, J.L., Rosenthal, J. Selective conversion of CO_2 to CO with high efficiency using an inexpensive bismuth-based electrocatalyst, J. Am. Chem. Soc. 2013, 135, 8798–8801.

[301] Ding, C., Li, A., Lu, S.-M., Zhang, H., Li, C. In situ electrodeposited indium nanocrystals for efficient CO_2 reduction to CO with low overpotential, ACS Catal. 2016, 6, 6438–6443.

[302] Medina-Ramos, J., DiMeglio, J.L., Rosenthal, J. Efficient reduction of CO_2 to CO with high current density using in situ or ex situ prepared Bi-based materials, J. Am. Chem. Soc. 2014, 136, 8361–8367.

[303] Zhu, Q., et al. Electrochemical reduction of CO_2 to CO using graphene oxide/carbon nanotube electrode in ionic liquid/acetonitrile system, Sci China Chem. 2016, 59, 551–556.

[304] Kumar, B., et al. Renewable and metal-free carbon nanofibre catalysts for carbon dioxide reduction, Nat. Commun. 2013, 4, 2819.

[305] Sun, X., et al. Very highly efficient reduction of CO_2 to CH_4 using metal-free N-doped carbon electrodes, Chem. Sci. 2016, 7, 2883–2887.

[306] Huan, T.N., et al. Porous dendritic copper: An electrocatalyst for highly selective CO_2 reduction to formate in water/ionic liquid electrolyte, Chem. Sci. 2017, 8, 742–747.

[307] Sun, X., et al. Design of a Cu(I)/C-doped boron nitride electrocatalyst for efficient conversion of CO_2 into acetic acid, Green Chem. 2017, 19, 2086–2091.

[308] Yang, D., et al. Selective electroreduction of carbon dioxide to methanol on copper selenide nanocatalysts, Nat Commun. 2019, 10, 677.

[309] Cadena, C., et al. Why is CO_2 so soluble in imidazolium-based ionic liquids?, J. Am. Chem. Soc. 2004, 126, 5300–5308.

[310] Lu, L., et al. Highly efficient electroreduction of CO_2 to methanol on palladium–copper bimetallic aerogels, Angew. Chem. Int. Ed. 2018, 57, 14149–14153.

[311] Asadi, M., et al. Robust carbon dioxide reduction on molybdenum disulphide edges, Nat Commun. 2014, 5, 4470.

[312] Mao, X., Wang, L., Xu, Y., Li, Y. Modulating the MoS_2 edge structures by doping transition metals for electrocatalytic CO_2 reduction, J. Phys. Chem. C 2020, 124, 10523–10529.

[313] Hong, X., Chan, K., Tsai, C., Nørskov, J.K. How doped MoS_2 breaks transition-metal scaling relations for CO_2 electrochemical reduction, ACS Catal. 2016, 6, 4428–4437.

[314] Lv, K., et al. Nitrogen doped MoS_2 and nitrogen doped carbon dots composite catalyst for electroreduction CO_2 to CO with high faradaic efficiency, Nano Energy 2019, 63, 103834.

[315] Abbasi, P., et al. Tailoring the edge structure of molybdenum disulfide toward electrocatalytic reduction of carbon dioxide, ACS Nano 2017, 11, 453–460.

[316] Sun, X., et al. Molybdenum–bismuth bimetallic chalcogenide nanosheets for highly efficient electrocatalytic reduction of carbon dioxide to methanol, Angew. Chem. Int. Ed. 2016, 55, 6771–6775.

[317] Wang, Y., et al. Copper nanocubes for CO_2 reduction in gas diffusion electrodes, Nano Lett. 2019, 19, 8461–8468.

[318] Yang, K.D., et al. Morphology-directed selective production of ethylene or ethane from CO_2 on a Cu mesopore electrode, Angew. Chem. Int. Ed. 2017, 56, 796–800.

[319] Ma, M., Djanashvili, K., Smith, W.A. Controllable hydrocarbon formation from the electrochemical reduction of CO_2 over Cu nanowire arrays, Angew. Chem. Int. Ed. 2016, 55, 6680–6684.

[320] Wu, M., et al. Promotion of CO_2 electrochemical reduction via Cu nanodendrites, ACS Appl. Mater. Interfaces 2020, 12, 11562–11569.

[321] Burdyny, T., et al. Nanomorphology-enhanced gas-evolution intensifies CO_2 reduction electrochemistry, ACS Sustain. Chem. Eng. 2017, 5, 4031–4040.

[322] Saberi Safaei, T., et al. High-density nanosharp microstructures enable efficient CO_2 electroreduction, Nano Lett. 2016, 16, 7224–7228.

[323] Chen, L.D., Urushihara, M., Chan, K., Nørskov, J.K. Electric field effects in electrochemical CO_2 reduction, ACS Catal. 2016, 6, 7133–7139.

[324] Liu, M., et al. Enhanced electrocatalytic CO_2 reduction via field-induced reagent concentration, Nature 2016, 537, 382–386.

[325] Higgins, D., Hahn, C., Xiang, C., Jaramillo, T.F., Weber, A.Z. Gas-diffusion electrodes for carbon dioxide reduction: A new paradigm, ACS Energy Lett. 2019, 4, 317–324.

[326] Omrani, R., Shabani, B. Gas diffusion layer modifications and treatments for improving the performance of proton exchange membrane fuel cells and electrolysers: A review, Int. J. Hydrogen Energy 2017, 42, 28515–28536.

[327] Lee, M., Huang, X. Development of a hydrophobic coating for the porous gas diffusion layer in a PEM-based electrochemical hydrogen pump to mitigate anode flooding, Electrochem. Commun. 2019, 100, 39–42.

[328] Xing, Z., Hu, L., Ripatti, D.S., Hu, X., Feng, X. Enhancing carbon dioxide gas-diffusion electrolysis by creating a hydrophobic catalyst microenvironment, Nat. Commun. 2021, 12, 136.

[329] Xu, A., et al. A facile solution to mature cathode modified by hydrophobic dimethyl silicon oil (DMS) layer for electro-fenton processes: Water proof and enhanced oxygen transport, Electrochim. Acta 2019, 308, 158–166.

[330] Li, J., et al. Efficient electrocatalytic CO_2 reduction on a three-phase interface, Nat. Catal. 2018, 1, 592–600.

[331] Firet, N.J., et al. Copper and silver gas diffusion electrodes performing CO_2 reduction studied through operando x-ray absorption spectroscopy, Catal. Sci. Technol. 2020, 10, 5870–5885.

[332] Burdyny, T., Smith, W.A. CO_2 reduction on gas-diffusion electrodes and why catalytic performance must be assessed at commercially-relevant conditions, Energy Environ. Sci. 2019, 12, 1442–1453.

[333] Zhang, J., Luo, W., Züttel, A. Self-supported copper-based gas diffusion electrodes for CO_2 electrochemical reduction, J. Mater. Chem. A 2019, 7, 26285–26292.

[334] Zahiri, B., Sow, P.K., Kung, C.H., Mérida, W. Active control over the wettability from superhydrophobic to superhydrophilic by electrochemically altering the oxidation state in a low voltage range, Adv. Mater. Interfaces 2017, 4, 1700121.

[335] Nesbitt, N.T., et al. Liquid–solid boundaries dominate activity of CO_2 reduction on gas-diffusion electrodes, ACS Catal. 2020, 10, 14093–14106.

[336] Junge Puring, K., et al. Electrochemical CO_2 reduction: Tailoring catalyst layers in gas diffusion electrodes, Adv. Sustain. Syst. 2021, 5, 2000088.

[337] Liang, X., et al. Ionomer cross-linking immobilization of catalyst nanoparticles for high performance alkaline membrane fuel cells, Chem. Mater. 2019, 31, 7812–7820.

[338] Pan, H., Barile, C.J. Electrochemical CO_2 reduction to methane with remarkably high faradaic efficiency in the presence of a proton permeable membrane, Energy Environ. Sci. 2020, 13, 3567–3578.

[339] García de Arquer, F.P., et al. CO_2 electrolysis to multicarbon products at activities greater than 1 A cm^{-2}, Science. 2020, 367, 661–666.

[340] Wakerley, D., et al. Bio-inspired hydrophobicity promotes CO_2 reduction on a Cu surface, Nat. Mater. 2019, 18, 1222–1227.

[341] Perry, S.C., et al. Polymers with intrinsic microporosity (PIMs) for targeted CO_2 reduction to ethylene, Chemosphere. 2020, 248, 125993.

[342] Xie, M.S., et al. Amino acid modified copper electrodes for the enhanced selective electroreduction of carbon dioxide towards hydrocarbons, Energy Environ. Sci. 2016, 9, 1687–1695.

[343] Kim, C., et al. Achieving selective and efficient electrocatalytic activity for CO_2 reduction using immobilized silver nanoparticles, J. Am. Chem. Soc. 2015, 137, 13844–13850.

[344] Han, Z., Kortlever, R., Chen, H.-Y., Peters, J.C., Agapie, T. CO_2 reduction selective for $C_{\geq 2}$ products on polycrystalline copper with N-substituted pyridinium additives, ACS Cent. Sci. 2017, 3, 853–859.

[345] Yang, H.-P., Qin, S., Wang, H., Lu, J.-X. Organically doped palladium: A highly efficient catalyst for electroreduction of CO_2 to methanol, Green Chem. 2015, 17, 5144–5148.

[346] Yang, H.-P., Yue, Y.-N., Qin, S., Wang, H., Lu, J.-X. Selective electrochemical reduction of CO_2 to different alcohol products by an organically doped alloy catalyst, Green Chem. 2016, 18, 3216–3220.

[347] Lee, J.H.Q., Lauw, S.J.L., Webster, R.D. The electrochemical reduction of carbon dioxide (CO_2) to methanol in the presence of pyridoxine (vitamin B_6), Electrochem. Commun. 2016, 64, 69–73.

[348] Giesbrecht, P.K., Herbert, D.E. Electrochemical reduction of carbon dioxide to methanol in the presence of benzannulated dihydropyridine additives, ACS Energy Lett. 2017, 2, 549–555.

[349] Lessio, M., Senftle, T.P., Carter, E.A. Is the surface playing a role during pyridine-catalyzed CO_2 reduction on p-GaP photoelectrodes?, ACS Energy Lett. 2016, 1, 464–468.

[350] Costentin, C., Drouet, S., Robert, M., Savéant, J.-M. A local proton source enhances CO_2 electroreduction to CO by a molecular Fe catalyst, Science 2012, 338, 90–94.

[351] Franco, F., et al. A local proton source in a [Mn(bpy-R)$(CO)_3$Br]-type redox catalyst enables CO_2 reduction even in the absence of Brønsted acids, Chem. Commun. 2014, 50, 14670–14673.

[352] Franco, F., et al. Local proton source in electrocatalytic CO_2 reduction with [Mn(bpy–R)$(CO)_3$Br] complexes, Chem. Eur. J. 2017, 23, 4782–4793.

[353] Zhou, Y., Xiao, Y., Zhao, J. A local proton source from carboxylic acid functionalized metal porphyrins for enhanced electrocatalytic CO_2 reduction, New J. Chem. 2020, 44, 16062–16068.

[354] Perry, S., et al. Hydrophobic thiol coatings to facilitate a triphasic interface for carbon dioxide reduction to ethylene at gas diffusion electrodes, Faraday Discuss. 2021. *Advance Article* DOI: 10.1039/D0FD00133C.

[355] Li, Y.C., et al. Bipolar membranes inhibit product crossover in CO_2 electrolysis cells, Adv. Sustain. Syst. 2018, 2, 1700187.

[356] Hagesteijn, K.F.L., Jiang, S., Ladewig, B.P. A review of the synthesis and characterization of anion exchange membranes, J. Mater. Sci. 2018, 53, 11131–11150.

[357] Li, Y.C., et al. Electrolysis of CO_2 to syngas in bipolar membrane-based electrochemical cells, ACS Energy Lett. 2016, 1, 1149–1153.

[358] Lin, M., Han, L., Singh, M.R., Xiang, C. An experimental- and simulation-based evaluation of the CO_2 utilization efficiency of aqueous-based electrochemical CO_2 reduction reactors with ion-selective membranes, ACS Appl. Energy Mater. 2019, 2, 5843–5850.

[359] Daud, S.S., Norrdin, M.A., Jaafar, J., Sudirman, R. The effect of material on bipolar membrane fuel cell performance: A review, IOP Conf. Ser.: Mater. Sci. Eng. 2020, 736, 032003.

[360] Liang, S., Altaf, N., Huang, L., Gao, Y., Wang, Q. Electrolytic cell design for electrochemical CO_2 reduction, J. CO2 Util. 2020, 35, 90–105.

[361] Hohenadel, A., et al. Electrochemical characterization of hydrocarbon bipolar membranes with varying junction morphology, ACS Appl. Energy Mater. 2019, 2, 6817–6824.

[362] Oener, S.Z., Foster, M.J., Boettcher, S.W. Accelerating water dissociation in bipolar membranes and for electrocatalysis, Science. 2020, 369, 1099–1103.

[363] Manohar, M., Shahi, V.K. Graphene oxide–polyaniline as a water dissociation catalyst in the interfacial layer of bipolar membrane for energy-saving production of carboxylic acids from carboxylates by electrodialysis, ACS Sustain. Chem. Eng. 2018, 6, 3463–3471.

[364] Hao, J.H., Chen, C., Li, L., Yu, L., Jiang, W. Preparation of bipolar membranes (I), J. Appl. Polym. Sci. 2001, 80, 1658–1663.

[365] Vermaas, D.A., Smith, W.A. Synergistic electrochemical CO_2 reduction and water oxidation with a bipolar membrane, ACS Energy Lett. 2016, 1, 1143–1148.

[366] Blommaert, M.A., Verdonk, J.A.H., Blommaert, H.C.B., Smith, W.A., Vermaas, D.A. Reduced ion crossover in bipolar membrane electrolysis via increased current density, molecular size, and valence, ACS Appl. Energy Mater. 2020, 3, 5804–5812.

[367] Shen, C., Wycisk, R., Pintauro, P.N. High performance electrospun bipolar membrane with a 3D junction, Energy Environ. Sci. 2017, 10, 1435–1442.

[368] Pătru, A., Binninger, T., Pribyl, B., Schmidt, T.J. Design principles of bipolar electrochemical Co-electrolysis cells for efficient reduction of carbon dioxide from gas phase at low temperature, J. Electrochem. Soc. 2019, 166, F34-F43.

[369] Yan, Z., Hitt, J.L., Zeng, Z., Hickner, M.A., Mallouk, T.E. Improving the efficiency of CO_2 electrolysis by using a bipolar membrane with a weak-acid cation exchange layer, Nat. Chem. 2021, 13, 33–40.

[370] Zhang, X., et al. Hierarchically ordered nanochannel array membrane reactor with three-dimensional electrocatalytic interfaces for electrohydrogenation of CO_2 to alcohol, ACS Energy Lett. 2018, 3, 2649–2655.

[371] Bisang, J.M. Wastewater treatment, electrochemical design concepts, in: Kreysa, G., Ota, K.-I., Savinell, R.F. (Eds.) Encyclopedia of applied electrochemistry, Springer New York, New York, 2014, 2132–2139.

[372] Vedharathinam, V., et al. Using a 3D porous flow-through electrode geometry for high-rate electrochemical reduction of CO_2 to CO in ionic liquid, ACS Catal. 2019, 9, 10605–10611.

[373] Weekes, D.M., Salvatore, D.A., Reyes, A., Huang, A., Berlinguette, C.P. Electrolytic CO_2 reduction in a flow cell, Acc. Chem. Res. 2018, 51, 910–918.

[374] Endrődi, B., et al. Continuous-flow electroreduction of carbon dioxide, Prog. Energy Combust. Sci. 2017, 62, 133–154.

[375] Yang, H., Kaczur, J.J., Sajjad, S.D., Masel, R.I. Performance and long-term stability of CO_2 conversion to formic acid using a three-compartment electrolyzer design, J. CO_2 Util. 2020, 42, 101349.

[376] Ma, S., et al. One-step electrosynthesis of ethylene and ethanol from CO_2 in an alkaline electrolyzer, J. Power Sources 2016, 301, 219–228.

[377] Kas, R., et al. Along the channel gradients impact on the spatioactivity of gas diffusion electrodes at high conversions during CO_2 electroreduction, ACS Sustain. Chem. Eng. 2021, 9, 1286–1296.

[378] Liu, H., Li, P., Juarez-Robles, D., Wang, K., Hernandez-Guerrero, A. Experimental study and comparison of various designs of gas flow fields to PEM fuel cells and cell stack performance, Front. Energy Res. 2014, 2, 2.

[379] Gundlapalli, R., Jayanti, S. Performance characteristics of several variants of interdigitated flow fields for flow battery applications, J. Power Sources 2020, 467, 228225.

[380] Manzi-Orezzoli, V., Siegwart, M., Cochet, M., Schmidt, T.J., Boillat, P. Improved water management for PEFC with interdigitated flow fields using modified gas diffusion layers, J. Electrochem. Soc. 2019, 167, 054503.

[381] Ripatti, D.S., Veltman, T.R., Kanan, M.W. Carbon monoxide gas diffusion electrolysis that produces concentrated C_2 products with high single-pass conversion, Joule. 2019, 3, 240–256.

[382] Nguyen, T.N., Dinh, C.-T. Gas diffusion electrode design for electrochemical carbon dioxide reduction, Chem. Soc. Rev. 2020, 49, 7488–7504.

[383] Garg, S., et al. Advances and challenges in electrochemical CO_2 reduction processes: An engineering and design perspective looking beyond new catalyst materials, J. Mater. Chem. A 2020, 8, 1511–1544.

[384] Lee, W., Kim, Y.E., Youn, M.H., Jeong, S.K., Park, K.T. Catholyte-free electrocatalytic CO_2 reduction to formate, Angew. Chem. Int. Ed. 2018, 57, 6883–6887.

[385] Fan, L., Xia, C., Zhu, P., Lu, Y., Wang, H. Electrochemical CO_2 reduction to high-concentration pure formic acid solutions in an all-solid-state reactor, Nat. Commun. 2020, 11, 3633.

[386] De Mot, B., Ramdin, M., Hereijgers, J., Vlugt, T.J.H., Breugelmans, T. Direct water injection in catholyte-free zero-gap carbon dioxide electrolyzers, ChemElectroChem 2020, 7, 3839–3843.

[387] Aeshala, L.M., Uppaluri, R.G., Verma, A. Effect of cationic and anionic solid polymer electrolyte on direct electrochemical reduction of gaseous CO_2 to fuel, J. CO_2 Util. 2013, 3–4, 49–55.

[388] Kutz, R.B., et al. Sustainion imidazolium-functionalized polymers for carbon dioxide electrolysis, Energy Technol. 2017, 5, 929–936.

[389] Yang, H., Kaczur, J.J., Sajjad, S.D., Masel, R.I. Electrochemical conversion of CO_2 to formic acid utilizing Sustainion™ membranes, J. CO2 Util. 2017, 20, 208–217.

[390] Chen, Y., et al. A robust, scalable platform for the electrochemical conversion of CO_2 to formate: Identifying pathways to higher energy efficiencies, ACS Energy Lett. 2020, 5, 1825–1833.

[391] Horii, D., Fuchigami, T., Atobe, M. A new approach to anodic substitution reaction using parallel laminar flow in a micro-flow reactor, J. Am. Chem. Soc. 2007, 129, 11692–11693.

[392] Monroe, M.M., Lobaccaro, P., Lum, Y., Ager, J.W. Membraneless laminar flow cell for electrocatalytic CO_2 reduction with liquid product separation, J. Phys. D: Appl. Phys. 2017, 50, 154006.

[393] Lu, X., Leung, D.Y.C., Wang, H., Maroto-Valer, M.M., Xuan, J. A pH-differential dual-electrolyte microfluidic electrochemical cells for CO_2 utilization, Renew. Energy 2016, 95, 277–285.

[394] Lu, X., Leung, D.Y.C., Wang, H., Xuan, J. A high performance dual electrolyte microfluidic reactor for the utilization of CO_2, Appl. Energy 2017, 194, 549–559.

[395] Agarwal, A.S., Rode, E., Sridhar, N., Hill, D. Conversion of CO_2 to value-added chemicals: Opportunities and challenges, in: Chen, W.-Y., Suzuki, T., Lackner, M. (Eds.) Handbook of climate change mitigation and adaptation, Springer, New York, 2014, 1–40.

[396] Lu, X., Leung, D.Y.C., Wang, H., Xuan, J. Characterization of a microfluidic reactor for CO_2 conversion with electrolyte recycling, Renew. Energy 2017, 102, 15–20.

[397] Gleede, B., Selt, M., Gütz, C., Stenglein, A., Waldvogel, S.R. Large, highly modular narrow-gap electrolytic flow cell and application in dehydrogenative cross-coupling of phenols, Org. Process Res. Dev. 2020, 24, 1916–1926.

[398] Plutschack, M.B., Pieber, B., Gilmore, K., Seeberger, P.H. The hitchhiker's guide to flow chemistry, Chem. Rev. 2017, 117, 11796–11893.

[399] Dufek, E.J., Lister, T.E., Stone, S.G., McIlwain, M.E. Operation of a pressurized system for continuous reduction of CO_2, J. Electrochem. Soc. 2012, 159, F514-F517.

[400] Hara, K., Kudo, A., Sakata, T. Electrochemical reduction of carbon dioxide under high pressure on various electrodes in an aqueous electrolyte, J. Electroanal. Chem. 1995, 391, 141–147.

[401] Kohjiro, H., Tadayoshi, S. Large current density CO_2 reduction under high pressure using gas diffusion electrodes, Bull. Chem. Soc. Jpn. 1997, 70, 571–576.

[402] Gabardo, C.M., et al. Combined high alkalinity and pressurization enable efficient CO_2 electroreduction to CO, Energy Environ. Sci. 2018, 11, 2531–2539.

[403] Wang, X., et al. Efficient methane electrosynthesis enabled by tuning local CO_2 availability, J. Am. Chem. Soc. 2020, 142, 3525–3531.

[404] Singh, M.R., Goodpaster, J.D., Weber, A.Z., Head-Gordon, M., Bell, A.T. Mechanistic insights into electrochemical reduction of CO_2 over Ag using density functional theory and transport models, Proc. Natl. Acad. Sci. 2017, 114, E8812-E8821.

[405] Moradzaman, M., Martínez, C.S., Mul, G. Effect of partial pressure on product selectivity in Cu-catalyzed electrochemical reduction of CO_2, Sustain. Energy Fuels 2020, 4, 5195–5202.

[406] Song, H., Song, J.T., Kim, B., Tan, Y.C., Oh, J. Activation of C_2H_4 reaction pathways in electrochemical CO_2 reduction under low CO_2 partial pressure, Appl. Catal. B 2020, 272, 119049.

[407] Ohya, S., Kaneco, S., Katsumata, H., Suzuki, T., Ohta, K. Electrochemical reduction of CO_2 in methanol with aid of CuO and Cu_2O, Catal. Today 2009, 148, 329–334.

[408] Hashiba, H., Yotsuhashi, S., Deguchi, M., Yamada, Y. Systematic analysis of electrochemical CO_2 reduction with various reaction parameters using combinatorial reactors, ACS Comb. Sci. 2016, 18, 203–208.

[409] Hashiba, H., et al. A broad parameter range for selective methane production with bicarbonate solution in electrochemical CO_2 reduction, Sustain. Energy Fuels 2017, 1, 1734–1739.

[410] Xiao, H., Cheng, T., Goddard, W.A., Sundararaman, R. Mechanistic explanation of the pH dependence and onset potentials for hydrocarbon products from electrochemical reduction of CO on Cu (111), J. Am. Chem. Soc. 2016, 138, 483–486.

[411] Löwe, A., et al. Influence of temperature on the performance of gas diffusion electrodes in the CO_2 reduction reaction, ChemElectroChem 2019, 6, 4497–4506.

[412] Leonard, M.E., Clarke, L.E., Forner-Cuenca, A., Brown, S.M., Brushett, F.R. Investigating electrode flooding in a flowing electrolyte, gas-fed carbon dioxide electrolyzer, ChemSusChem 2020, 13, 400–411.

[413] Endrődi, B., et al. Multilayer electrolyzer stack converts carbon dioxide to gas products at high pressure with high efficiency, ACS Energy Lett. 2019, 4, 1770–1777.

[414] Jännsch, Y., et al. Pulsed potential electrochemical CO_2 reduction for enhanced stability and catalyst reactivation of copper electrodes, Electrochem. Commun. 2020, 121, 106861.

[415] Zhou, F., Li, H., Fournier, M., MacFarlane, D.R. Electrocatalytic CO_2 reduction to formate at low overpotentials on electrodeposited Pd films: Stabilized performance by suppression of CO formation, ChemSusChem 2017, 10, 1509–1516.

[416] Lee, C.W., Cho, N.H., Nam, K.T., Hwang, Y.J., Min, B.K. Cyclic two-step electrolysis for stable electrochemical conversion of carbon dioxide to formate, Nat. Commun. 2019, 10, 3919.

[417] Wuttig, A., Surendranath, Y. Impurity ion complexation enhances carbon dioxide reduction catalysis, ACS Catal. 2015, 5, 4479–4484.

[418] Seh, Z.W., et al. Combining theory and experiment in electrocatalysis: Insights into materials design, Science 2017, 355, eaad4998.

[419] Suen, N.-T., et al. Electrocatalysis for the oxygen evolution reaction: Recent development and future perspectives, Chem. Soc. Rev. 2017, 46, 337–365.

[420] Xia, C., Xia, Y., Zhu, P., Fan, L., Wang, H. Direct electrosynthesis of pure aqueous H_2O_2 solutions up to 20% by weight using a solid electrolyte, Science 2019, 366, 226–231.

[421] Glenk, G., Reichelstein, S. Economics of converting renewable power to hydrogen, Nat. Energy 2019, 4, 216–222.

[422] Hou, M., et al. A clean and membrane-free chlor-alkali process with decoupled Cl_2 and H_2/NaOH production, Nat Commun. 2018, 9, 438.

[423] Martínez, N.P., Isaacs, M., Nanda, K.K. Paired electrolysis for simultaneous generation of synthetic fuels and chemicals, New J. Chem. 2020, 44, 5617–5637.

[424] Gomez, E., Yan, B., Kattel, S., Chen, J.G. Carbon dioxide reduction in tandem with light-alkane dehydrogenation, Nat. Rev. Chem. 2019, 3, 638–649.

[425] Fuku, K., et al. Photoelectrochemical hydrogen peroxide production from water on a WO_3/$BiVO_4$ photoanode and from O_2 on an Au cathode without external bias, Chem. Asian J. 2017, 12, 1111–1119.

[426] Perry, S.C., et al. Electrochemical synthesis of hydrogen peroxide from water and oxygen, Nat. Rev. Chem. 2019, 3, 442–458.

[427] Sayama, K. Production of high-value-added chemicals on oxide semiconductor photoanodes under visible light for solar chemical-conversion processes, ACS Energy Lett. 2018, 3, 1093–1101.

[428] Vass, Á., Endrődi, B., Janáky, C. Coupling electrochemical carbon dioxide conversion with value-added anode processes: An emerging paradigm, Curr. Opin. Electrochem. 2021, 25, 100621.

[429] Llorente, M.J., Nguyen, B.H., Kubiak, C.P., Moeller, K.D. Paired electrolysis in the simultaneous production of synthetic intermediates and substrates, J. Am. Chem. Soc. 2016, 138, 15110–15113.

[430] Pérez-Gallent, E., et al. Electroreduction of CO_2 to CO paired with 1,2-propanediol oxidation to lactic acid. Toward an economically feasible system, Ind. Eng. Chem. Res. 2019, 58, 6195–6202.

[431] Vo, T., et al. Formate: An energy storage and transport bridge between carbon dioxide and a formate fuel cell in a single device, ChemSusChem 2015, 8, 3853–3858.

[432] Bienen, F., et al. Utilizing formate as an energy carrier by coupling CO_2 electrolysis with fuel cell devices, Chem. Ing. Tech. 2019, 91, 872–882.

[433] An, L., Chen, R. Direct formate fuel cells: A review, J. Power Sources 2016, 320, 127–139.

[434] Xiang, H., et al. Production of formate by CO_2 electrochemical reduction and its application in energy storage, Sustain. Energy Fuels 2020, 4, 277–284.

[435] Xia, C., et al. Continuous production of pure liquid fuel solutions via electrocatalytic CO_2 reduction using solid-electrolyte devices, Nat. Energy 2019, 4, 776–785.

[436] Ong, B.C., Kamarudin, S.K., Basri, S. Direct liquid fuel cells: A review, Int. J. Hydrogen Energy 2017, 42, 10142–10157.

[437] Pu, T., Tian, H., Ford, M.E., Rangarajan, S., Wachs, I.E. Overview of selective oxidation of ethylene to ethylene oxide by Ag catalysts, ACS Catal. 2019, 9, 10727–10750.

[438] van Bavel, S., Verma, S., Negro, E., Bracht, M. Integrating CO_2 electrolysis into the gas-to-liquids–power-to-liquids process, ACS Energy Lett. 2020, 5, 2597–2601.

[439] Zhou, X., Xiang, C. Comparative analysis of solar-to-fuel conversion efficiency: A direct, one-step electrochemical CO_2 reduction reactor versus a two-step, cascade electrochemical CO_2 reduction reactor, ACS Energy Lett. 2018, 3, 1892–1897.

[440] Creel, E.B., McCloskey, B.D. Scalable CO_2-to-oxygenate production, Nat. Catal. 2018, 1, 6–7.

[441] Jensen, M.T., et al. Scalable carbon dioxide electroreduction coupled to carbonylation chemistry, Nat. Commun. 2017, 8, 489.

[442] Zhong, S., et al. Electrochemical conversion of CO_2 to 2-bromoethanol in a membraneless cell, ACS Energy Lett. 2019, 4, 600–605.

Index

https://doi.org/10.1515/9781501522239-008

www.ingramcontent.com/pod-product-compliance
Lightning Source LLC
Chambersburg PA
CBHW081538220326

41598CB00036B/6473